Table of Contents

Chapter 1 : Rigging   4

Chapter 2 : Hydraulic Systems   47

Chapter 3 : Trim Tabs   100

Marine Auxiliary Systems

Copyright @ 2012
by
Marine Technical Training

By: Alvaro Lopez

Edition 2014

Marine Technical Training

It is a company dedicated to training Engineers and Marine Technicians in areas such as

- **Marine Electricity**
- **Marine Electronics**
- **Air Conditioning**
- **Diesel Engines**
- **Gasoline Engines**
- **Auxiliary Systems**
  - Hydraulics
  - Steering Systems
  - Trim Tabs
  - Bow Thrusters
  - Water makers

Copyrights

ABYC Certification

This book was designed like a consult handbook for Service, wiring, installation and diagnosis of typical auxiliary equipment in pleasure yachts. Also, it can be used as a textbook for marine engineering courses and as a complementary study guide in order to take the **ABYC Systems** and ABYC **Air Conditioning Certifications**

What are Auxiliary Systems

- Except for propulsion and steering systems , the rest of equipment and systems installed in a pleasure yacht are considered Auxiliary Equipment
- With the new electronics advances , the auxiliary equipment add more comfort and value to the yachting industry
- For this reason a marine technician specialized in auxiliary equipment require good knowledge in: Nautical Materials , Electricity , Electronics and Hydraulic Systems

Auxiliary Systems Classification

- The Auxiliary Systems are classified in seven basic categories
    1. Hydraulic & Pneumatic Systems
    2. Stabilizers and level control systems
    3. Steering Systems
    4. Thrusters and Anchor Systems
    5. Freshwater systems ,Potable water and Sanitation systems
    6. Air Conditioning and Refrigeration Systems
    7. Fire Protection Systems

Chapter 1
Rigging

- Navigation Lights
- Wire Ropes & Synthetic Ropes
- Pulleys
- Engine Controls
- Sea Suction
- Under Water Lights

Rigging

- Rigging , can be defined as the system of ropes, chains, and tackle used to support and control the masts, sails, and yards of a sailing vessel

- In mechanical terms this term can be used to study the different types of controls and mechanisms used to operate properly an hydraulic or electro mechanical system

Navigation Lights Rules

Port side lights are red , Starboard side lights are green and shine from dead ahead to 112.5° aft on either side

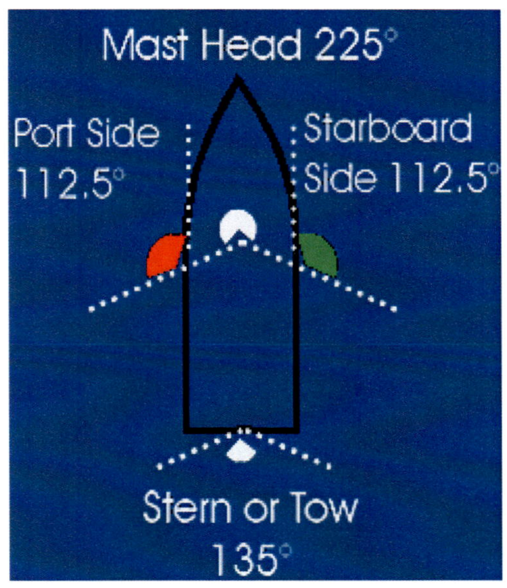

Navigation Lights Rules

Stern lights are white and shine aft and 67.5° forward on each side

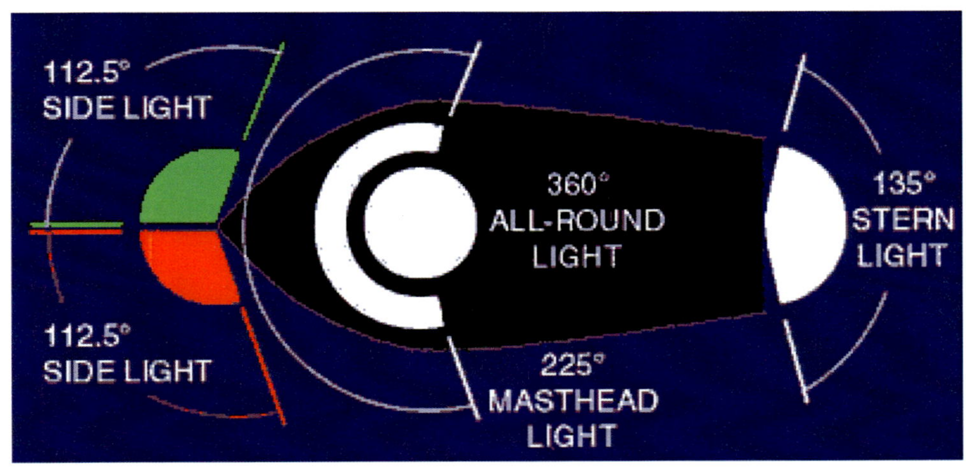

Navigation Lights Rules

All-round lights are white and shine through 360°

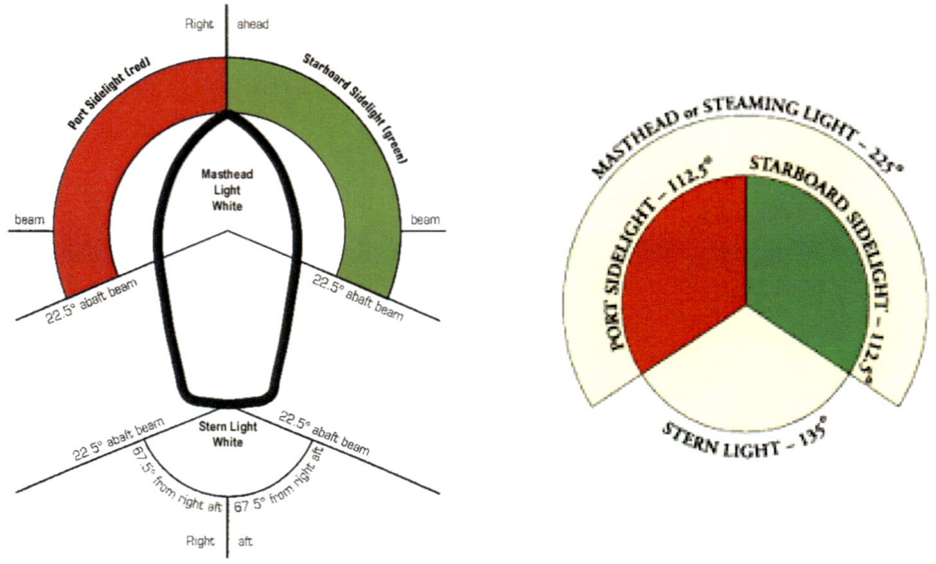

Masthead Lights Rules

Masthead lights are white and shine from 112.5° on the port side through dead ahead to 112.5° on the starboard side. They must be above the side lights

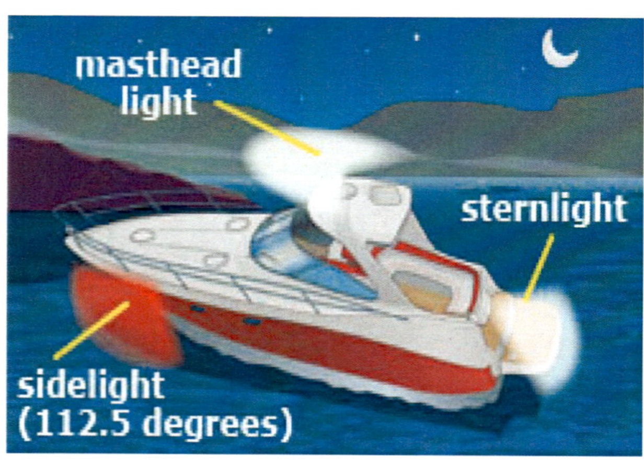

Navigation Lights Rules

Powerboats less than 20m (65.5') in length need to show side lights, a stern light and a masthead light. Power vessels less than 12m may show a single all-round light in lieu of the separate masthead and stern lights

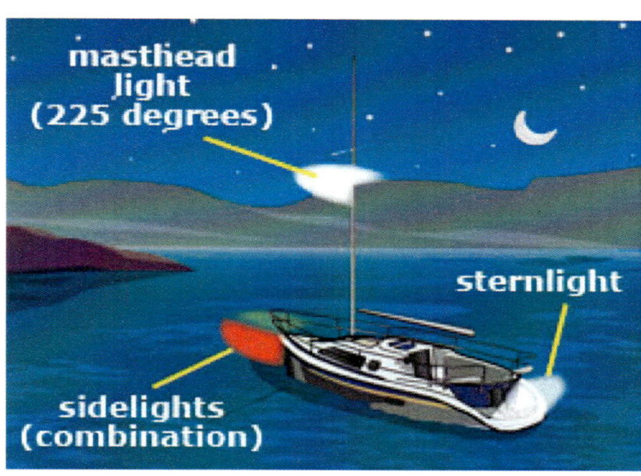

Navigation Lights Rules

Sailing vessels less than 20m in length need to show side lights and a stern light. These may be combined into a bicolor light and stern light, or a single tricolor light at the top of the mast

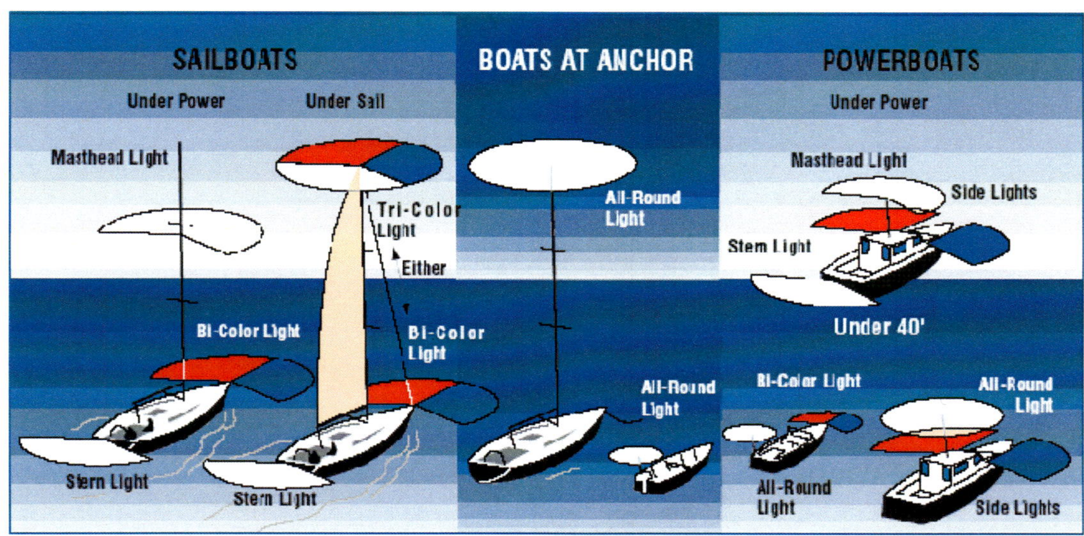

Nav. Lights Control Panel

The navigation lights control panel is located at the pilot helm console

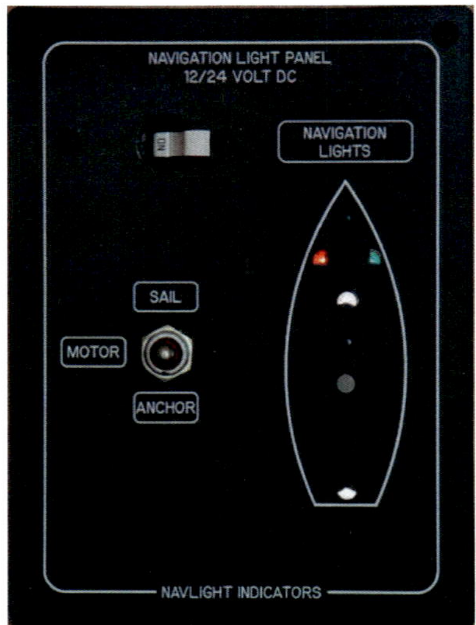

Nav Lights Wiring

A good practice is feed the navigation lights circuit breaker directly from the house battery bank

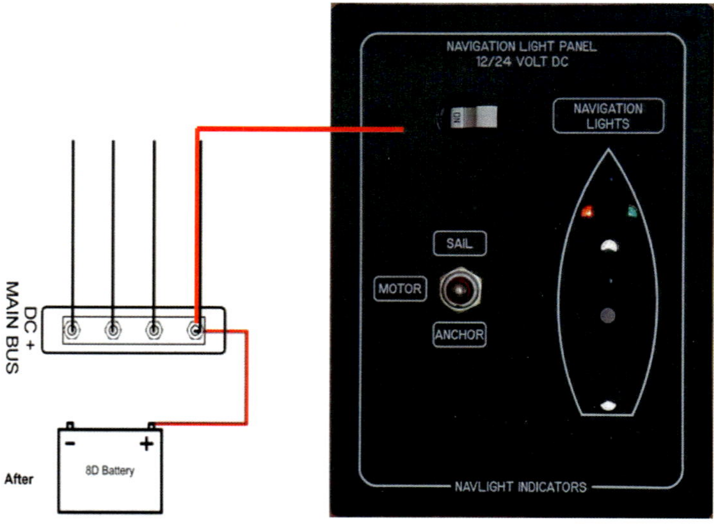

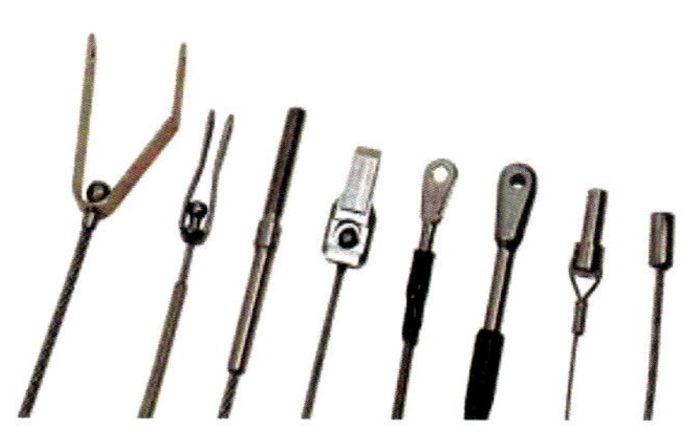 

Wire ropes and Synthetic ropes

- Wire Ropes
- Cable Assemblies
- Terminations
- Synthetic ropes

Wire Rope

- **Wire rope** is a type of <u>rope </u>which consists of several strands of metal wire laid (or 'twisted') into a <u>helix</u>

- Historically wire rope evolved from steel chains which had a record of mechanical failure. While flaws in chain links or solid steel bars can lead to catastrophic failure, flaws in the wires making up a steel cable are less critical as the other wires easily take up the load

Wire Rope Construction

Spiral ropes . In the so-called cross lay strands, the wires of the different layers cross each other. In the mostly used parallel lay strands, the lay length of all the wire layers is equal and the wires of any two superimposed layers are parallel, resulting in linear contact

Wire Rope Construction

Stranded ropes . In principle, spiral ropes are round strands as they have an assembly of layers of wires laid helically over a center with at least one layer of wires being laid in the opposite direction to that of the outer layer

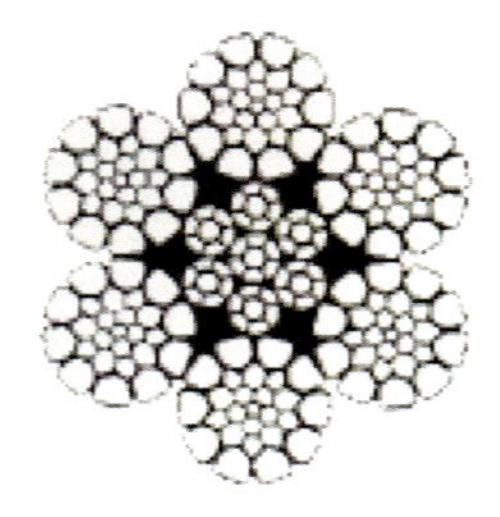
6 x 26 WARRINGTON SEALE
COMPACTED STRAND
IWRC

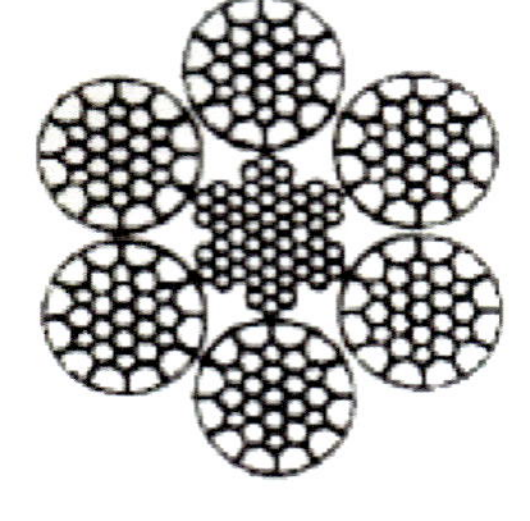
6 x 31 (1/6/6+6/12)
WARRINGTON-SEALE

19 x 19 SEALE
COMPACTED STRAND

Terminations

There are different ways of securing the ends of wire ropes to prevent fraying. The most common and useful type of end fitting for a wire rope is to turn the end back to form a loop

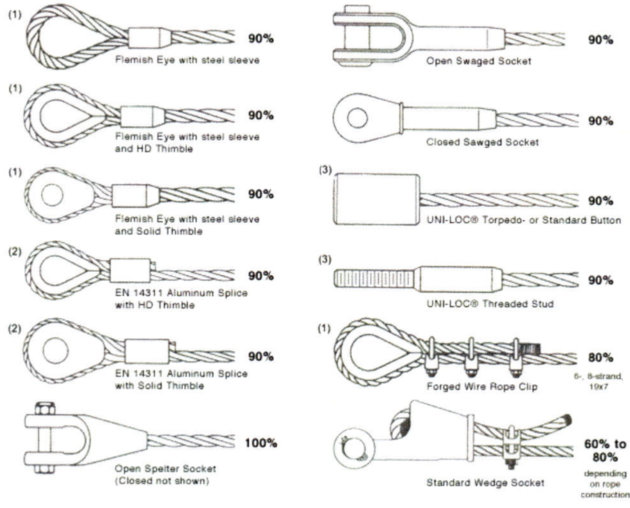

Cable Assemblies

Nylon Ropes

The nylon ropes has great strength, specially under load to absorb energy, It's also easy to handle and resists the effects of sunlight better than other synthetics

The nylon ropes has a good flexibility and also absorb and release humidity easily

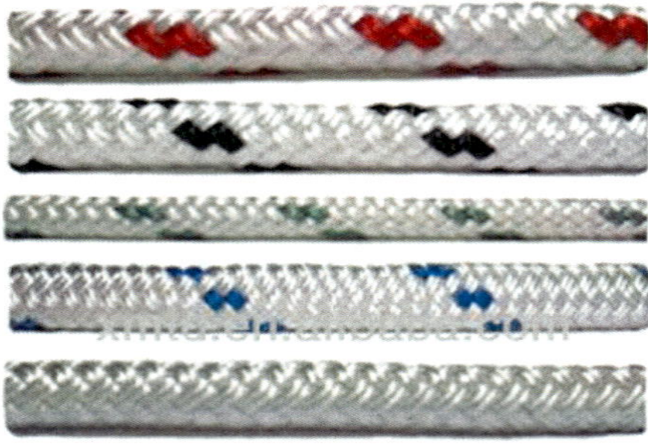

Nylon Ropes

- The Nylon ropes comes in strands and braided

- Three strand is usually used on anchor rodes because of its stretch and resistance to abrasion

- Braided, more commonly seen on dock lines and in sailing rigging, will snag easier than stranded line

12

Polypropylene Ropes

This the rope know as the yellow cord that's commonly used to tow skiers. Because polypropylene rope floats it's used for multiple uses

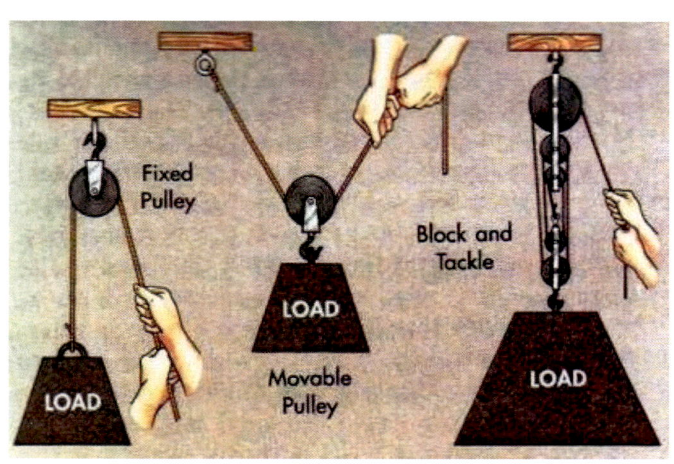

Pulleys

- Types of Pulleys
- Mechanical Advantage Calculation
- System of Pulleys
- Uses and Applications

Types of Pulleys

Fixed: A *fixed* pulley has an axle mounted in bearings attached to a supporting structure. A fixed pulley changes the direction of the force on a rope or belt that moves along its circumference

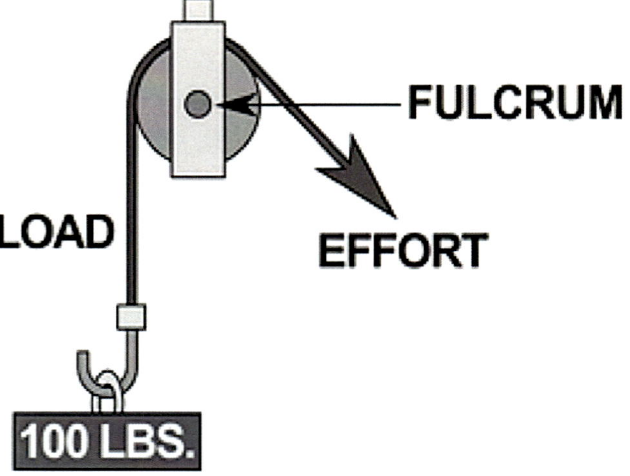

Types of Pulleys

Movable: A *movable* pulley has an axle in a movable block. A single movable pulley is supported by two parts of the same rope and has a mechanical advantage of two

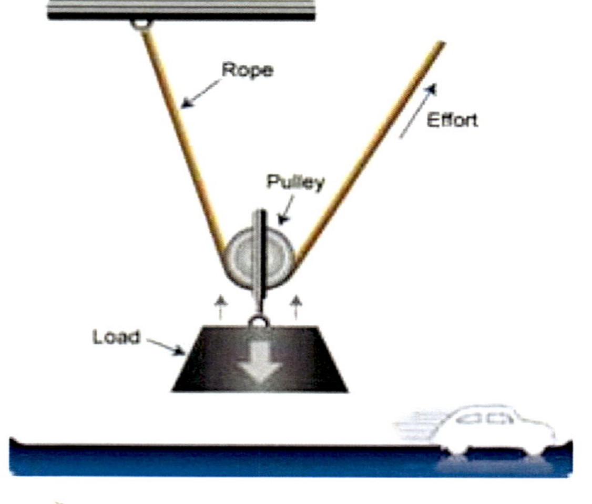

Types of Pulleys

Compound: A combination of fixed and a movable pulleys forms a <u>block and tackle</u>. A *block and tackle* can have several pulleys mounted on the fixed and moving axles, further increasing the mechanical advantage

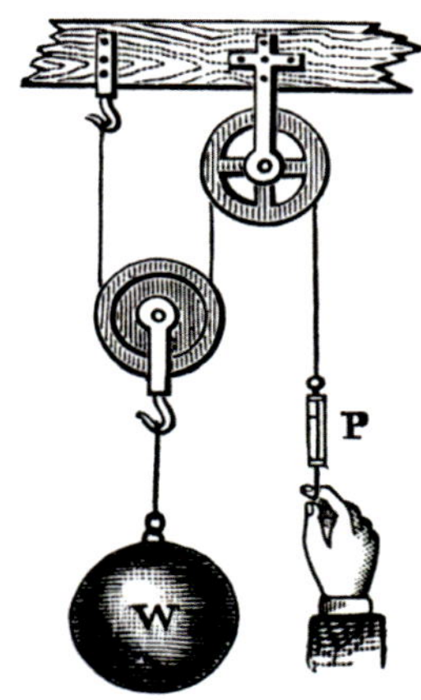

Block and tackle

A set of pulleys assembled so they rotate independently on the same axle form a block. Two blocks with a rope attached to one of the blocks and threaded through the two sets of pulleys form a <u>block and tackle</u>

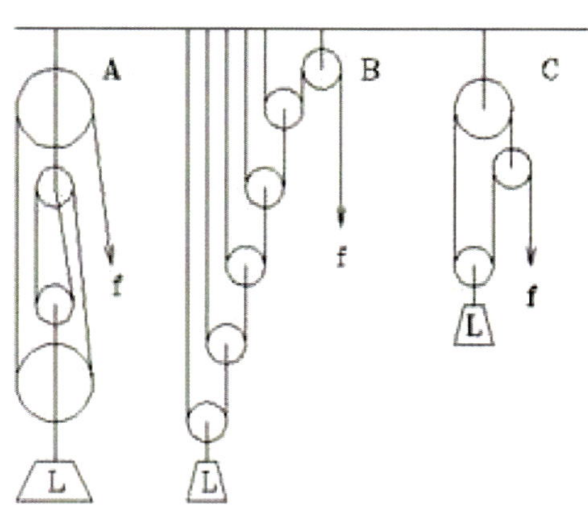

Mechanical Advantage of Pulleys

- A single pulley is used to change the direction of the pull to lift a load
- A system of pulleys is used to lift heavy loads by the **mechanical advantage (MA)** of increasing the length of travel of the pull and decreasing the load

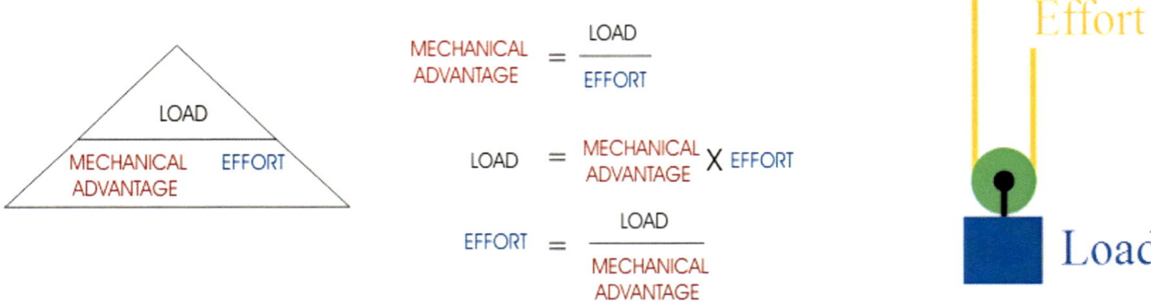

Mechanical Advantage Calculation

- Count the number of rope segments on each side of the pulleys, including the free end .
- If the free end is to be pulled down, subtract 1 from this number. This number is the mechanical advantage of the system

P = Efford

W = Load

MA =W/P = 2

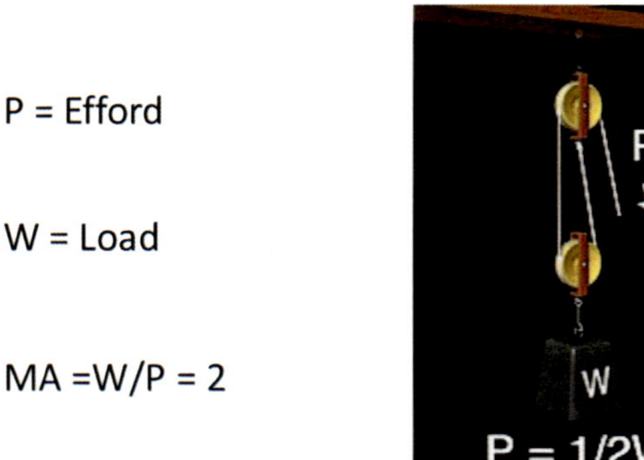

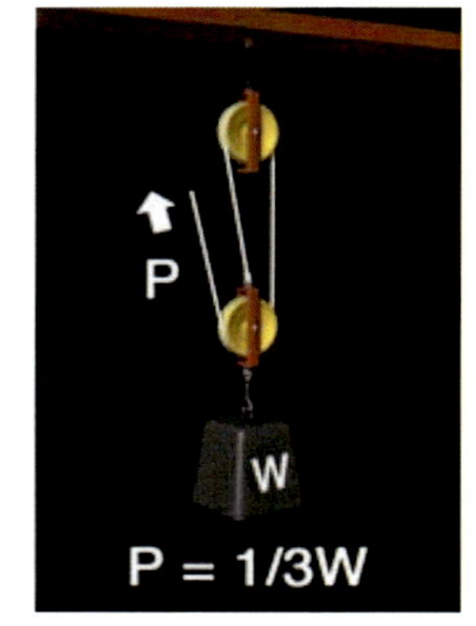

Mechanical Advantage Calculation

- To compute the amount of force necessary to hold the weight in equilibrium, divide the weight by the mechanical advantage! In the figure, for example, there are 3 sections of rope. Since the applied force is downward, we subtract 1 for a mechanical advantage of 2. It will take a force equal to 1/2 the weight to hold the weight steady

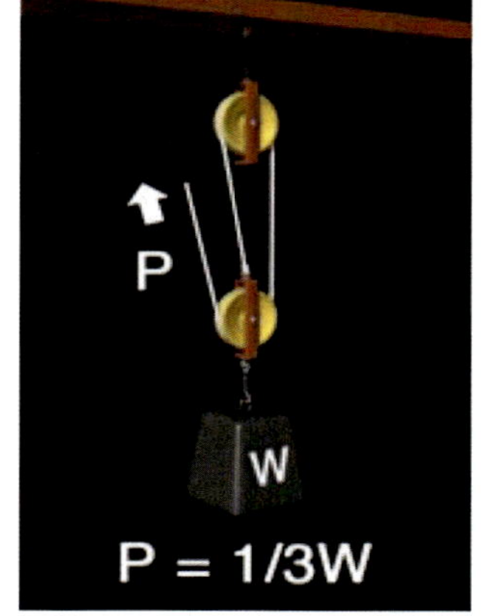

Mechanical Advantage Calculation

The next figure has the same two pulleys, but the rope is applied differently and it is pulled upwards. The mechanical advantage is 3, and the force to hold the weight in equilibrium is 1/3 the weight

Mechanical Advantage Calculation

Each additional figure shows another possible pulley configuration and lists the force necessary to lift and hold the weight still

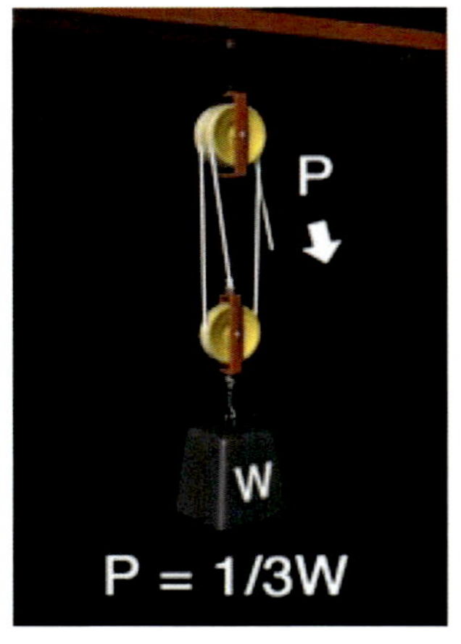

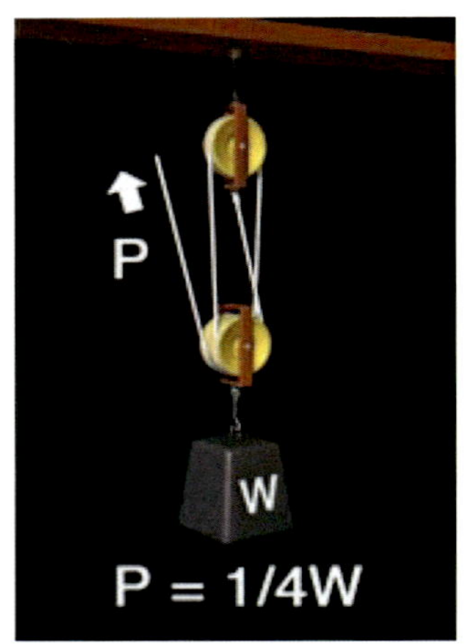

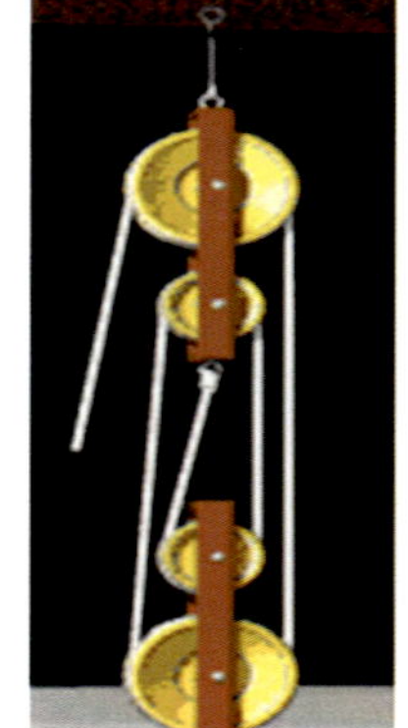

This rope and pulley system supports the 400 lb weight. How much tension in the rope is needed to hold the weight in equilibrium.

(Check the correct answer.)

A.- 100 lb.

B.- 80 lb

C.- 44.44 lb

This rope and pulley system supports a 75 lb weight. How much tension in the rope is needed to hold the weight in equilibrium

(Check the correct answer)

37.5 lbs
25 lbs
21 lbs

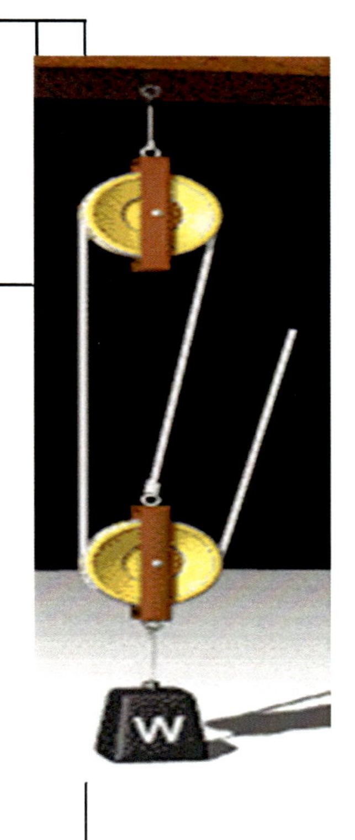

How much force does the person need to pull with so that the scale will read 100 lbs.

(Check the correct answer.)

A.- 75 lb
B.- 66.7 lb
C.- 80 lb

Engine Controls

- Mechanical System
- Minimum Bending radius
- Control Cable Routing
- Control cables Lubrication
- Electronic System
- Conversion From Mechanical to Electronic
- Hydraulic System

Mechanical Controls

Due to its lack of accuracy and their low durability, these systems are less used every day

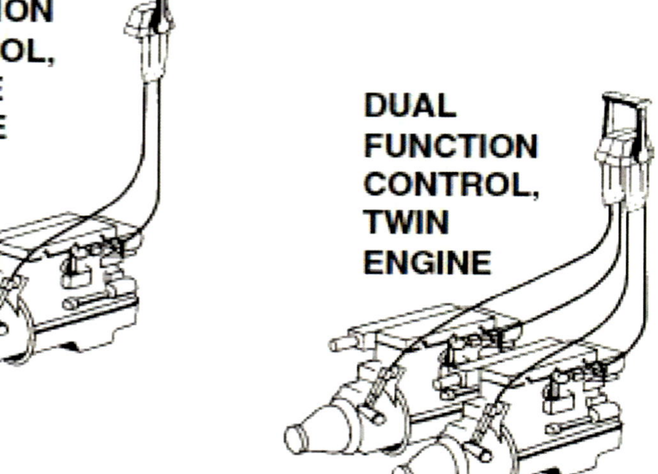

Mechanical System Uses

- They are commonly used in the systems such as:
  - Throttle and Propulsion systems
  - Shut off Valves

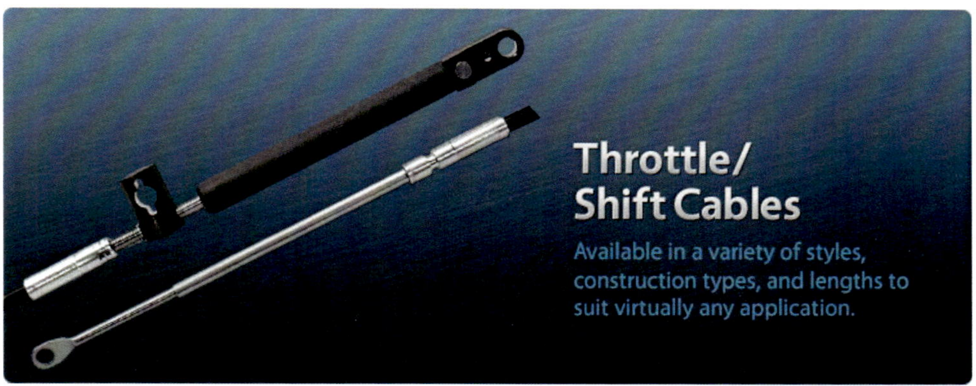

Control Cable Components

A good installation includes a clear route avoiding short bend radius and water contact

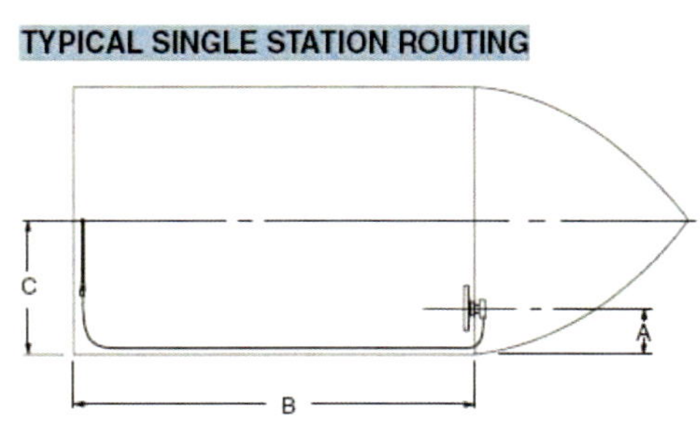

Minimum Bending Radius

- **Standard cables** use stainless steel Lubri-Core core wire for smooth operation and long life. **(8" minimum bend radius.)**
- **Midrange cables** use a heavier jacket and either stainless steel Lubri-Core or a coated cable (33C Supreme) core wire. (8" minimum bend radius for Lubri-Core construction, 5" for Supreme construction)
- **Premium cables** have a heavy jacket and the **TFXTREME** splined, coated core element for maximum smoothness with minimal lost motion. (4" min. bend radius.)

Cable Control Routing

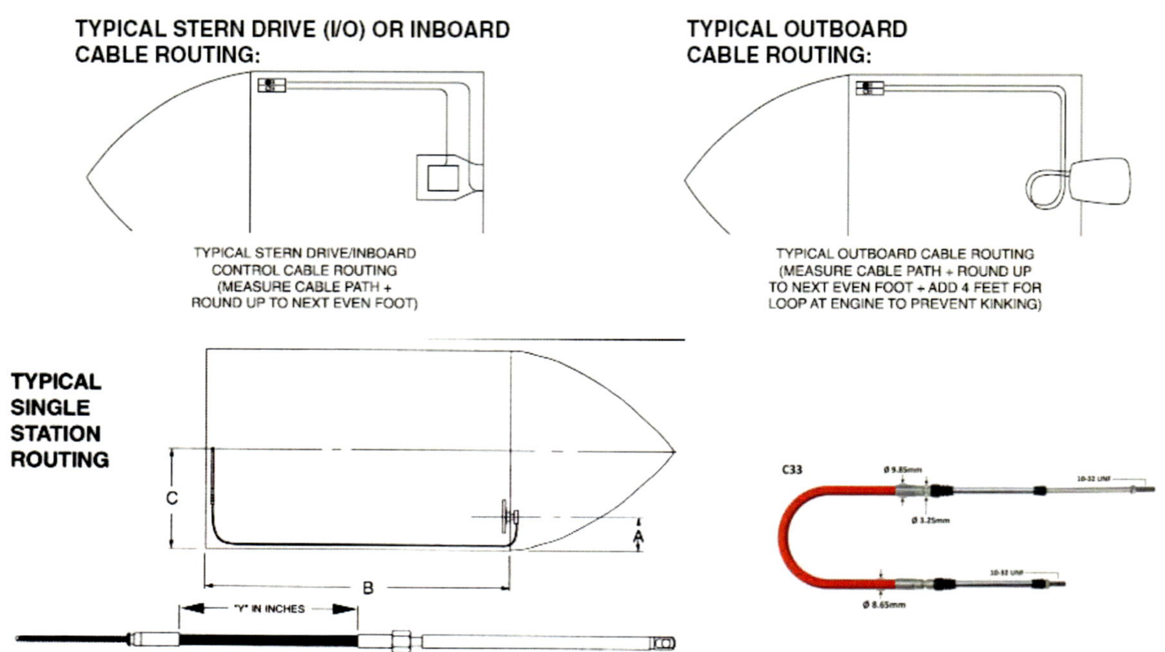

Control Cable Adjustment

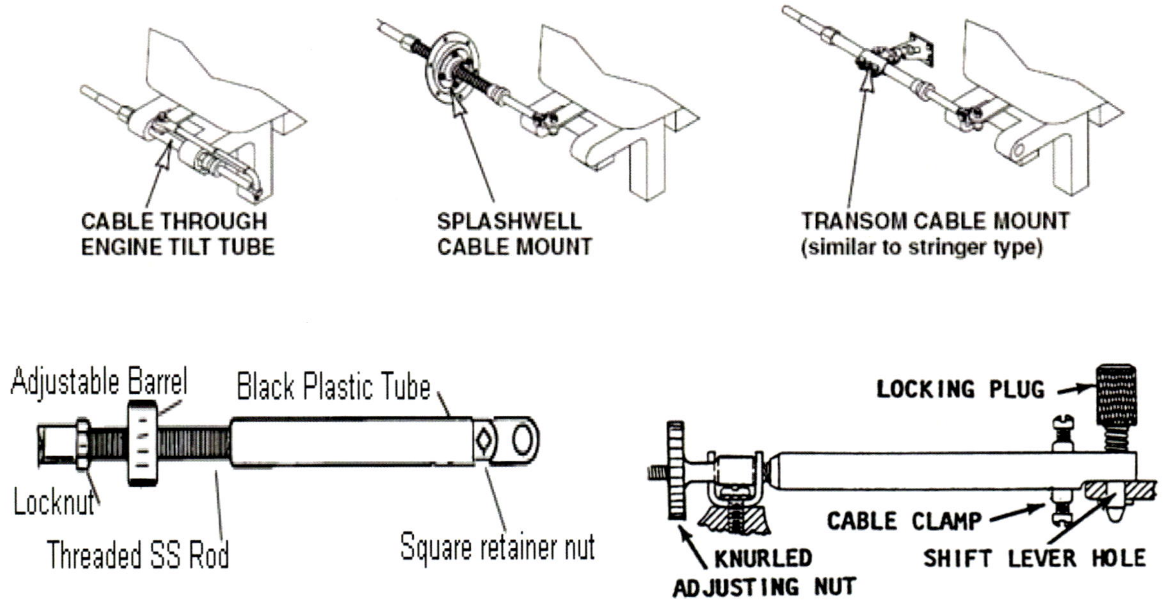

Control Cables Lubrication

Lubricants applied to the ends of the cable tend to penetrate only a short distance, often failing to deliver any meaningful benefit

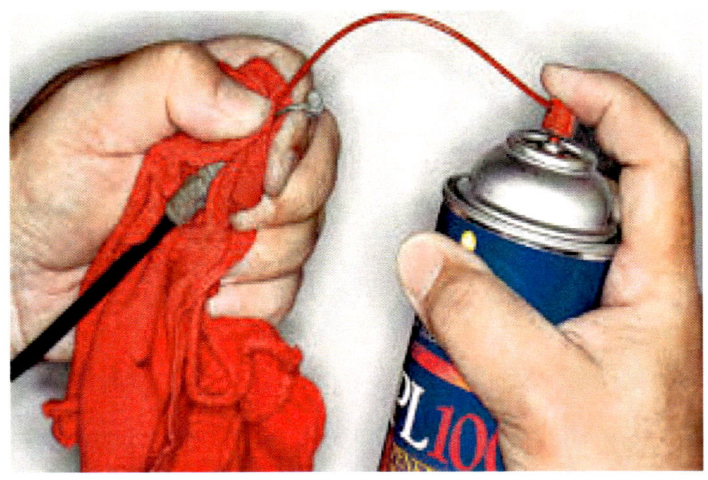

Control Cables Lubrication

To lubricate the entire cable, you must remove it from the boat. This can be dead simple or an all-day job, depending on how the cables are installed and routed on your boat

Control cables that are really difficult to remove and install make a strong case for replacement rather than the band-aid repair of lubrication

If your cables are accessible, lubricating can be an expenditure-deferring alternative

Lubrication Procedure

With the cable out of the boat, insert one end (jacket and all) through a hole you make in a bottom corner of a heavy-duty zip-seal bag

Lubrication Procedure

Place a container beneath the bottom end of the cable to catch the oil that should eventually drip out of the lower end of the jacket. Allow the cable to hang until the oil drains through

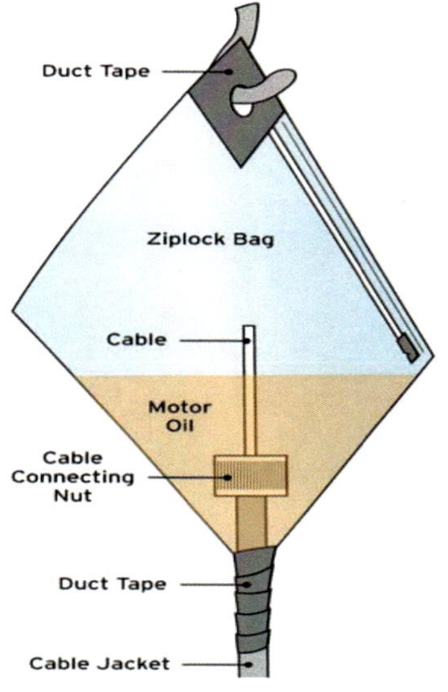

Electronic Controls

These controls are ranked: <u>wired and wireless</u>. The wired systems work together with some transducers and solenoids that convert electrical signal into mechanical movements

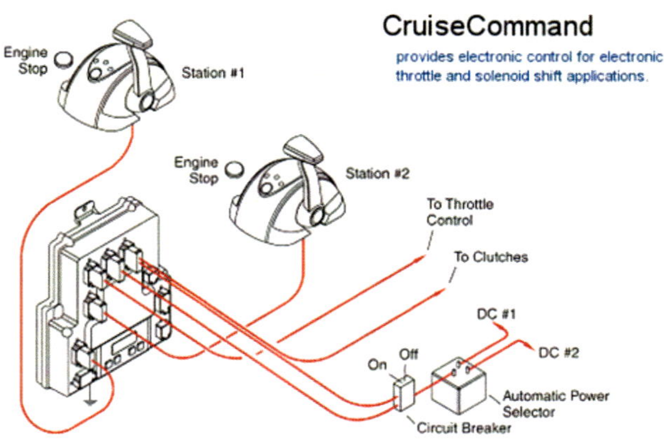

Electronic Controls

The wireless systems are in continuous development. The last versions require a constant internet connection to keep the system updated with the latest software releases

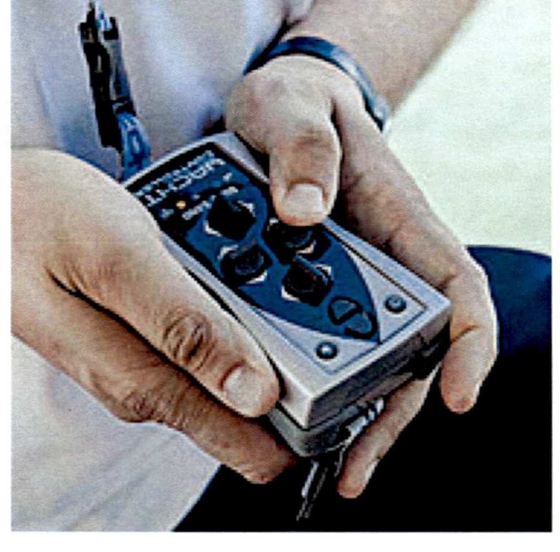

From Mechanical to Electronic

For boats with mechanical throttle and mechanical shift, the KE-4+ is a perfect answer to smooth and easy shift and throttle control suitable for most applications

- Up to 4 stations
  - Synchronization standard, 1 lever/2 lever selection

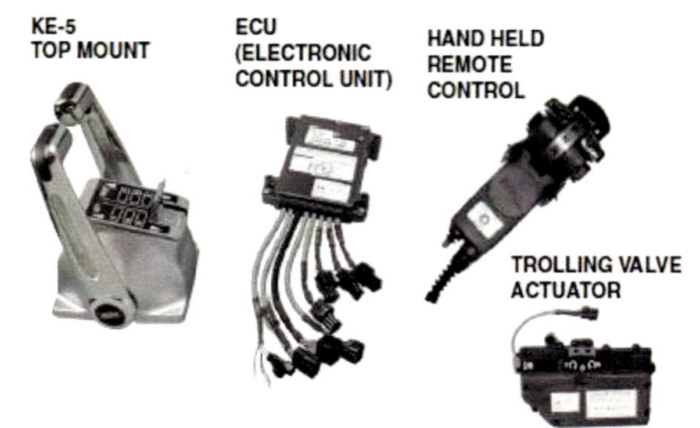

Electronic Controls

Electronic engine controls replace the standard helm unit and mechanical cables running from the helm to the engine room

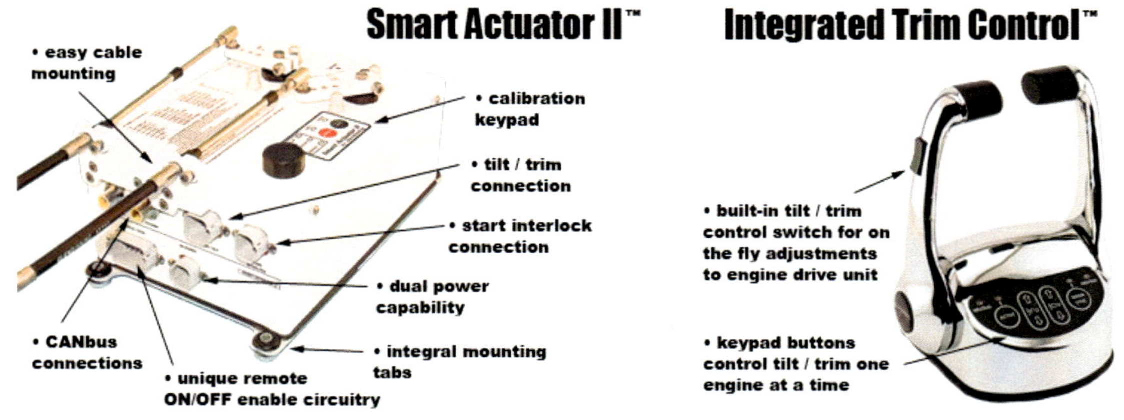

Electronic Controls

The electronic controls ,simplifies wiring and allows greater flexibility when adding control stations by utilizing a single communication cable

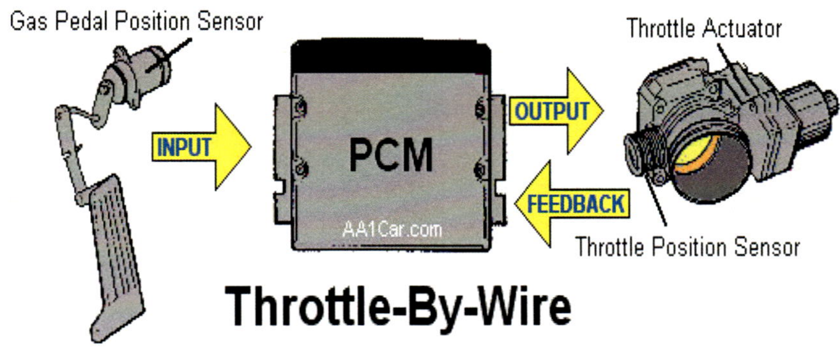

Throttle-By-Wire

Electronic Controls

Plug-n-Play Installation - deutsch connectors make for easier installation and more secure connection of cables minimizing wiring connection problems

Multiple Engine Applications

The Smart Actuator II system provides complete control of one, two, or three engines - inboard and outboards!

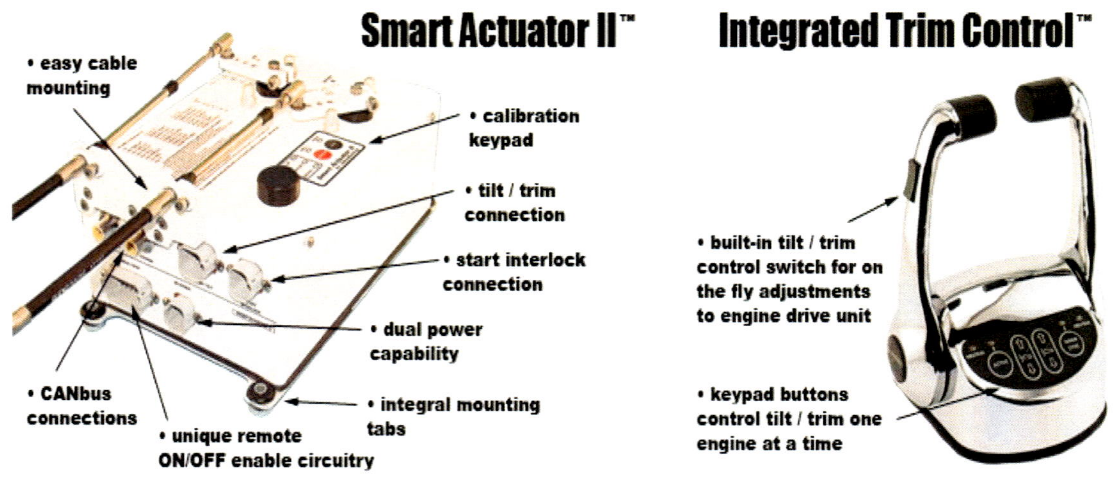

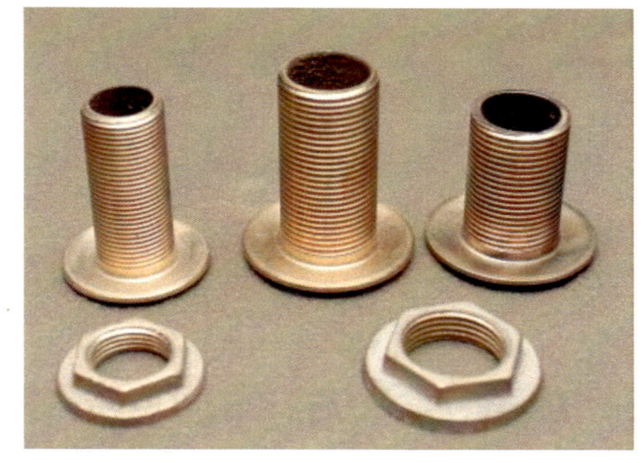

Sea Water Suction

- Through-Holes
- Seacocks
- strainers
- Heat-exchangers

Sea Suction

- Sea suction requires an opening in the Hull below the water line
- The sea suction has to be set up to provide adequate water supply, to not clog and to be properly inspected and serviced

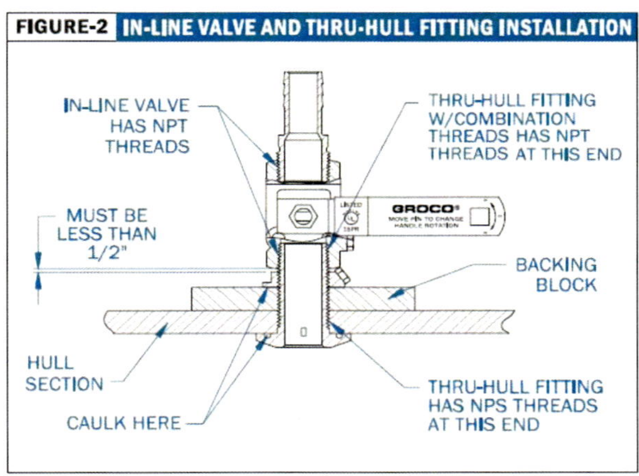

Sea Suction

- The fitting that penetrates the hull is called a through-hull

- On metal vessels, it may be a flanged valve bolted on a metal "spool" instead

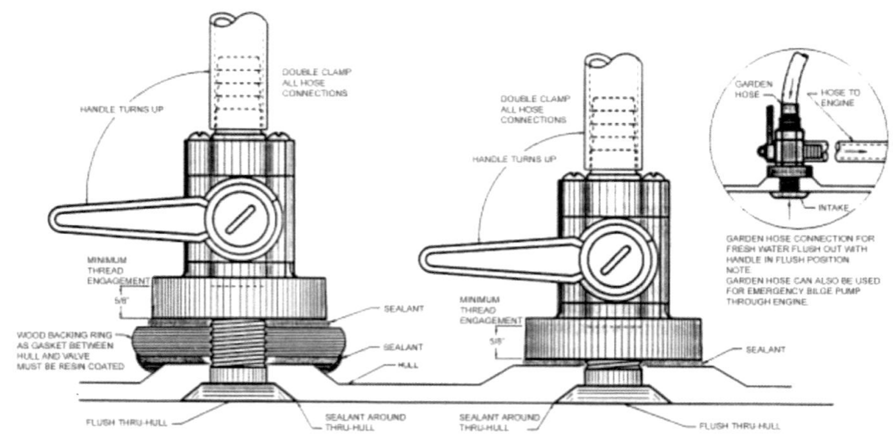

New Through-Hull

- Always locate **sea-cocks** (Transducers) where they are readily accessible or you defeat the purpose of having a valve in the line

- Before you drill the hole, double-check the location carefully both inside and outside the hull to make sure, for example, that inside there will be ample room to throw the handle, and outside the new fitting is not going to set up turbulence in front of your depth sounder or speed log impeller

New Through-Hull

- From inside the hull, drill a small pilot hole and check the location one more time.
- Protect the external surface (Gel-coat) with blue tape

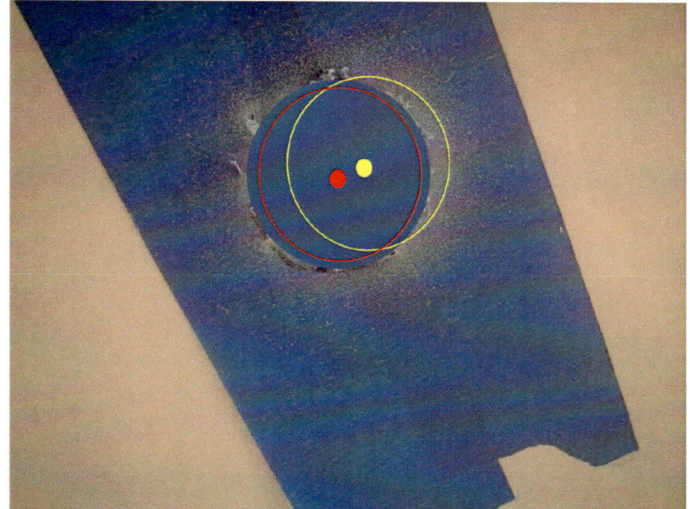

New Through-Hull

- Select a hole saw the size of the fitting you are installing and cut the required hole by first drilling from the outside of the hull until the pilot drill in the hole saw penetrates the hull, then finish the hole by drilling from the inside. Clean up the edges of the hole with emery cloth

Cored-Hull

- If the hull is cored, dig out the core around the hole to hollow an area at least as large as the flange of the seacock you are installing

- Fill the hollow area with epoxy putty and allow it to fully harden before proceeding .

- The epoxy provides a solid base for the through-hull and prevents water from reaching the core material

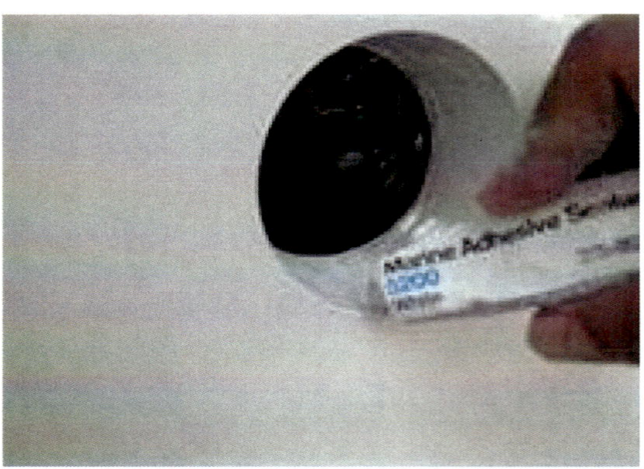

Dry-Fit

Cut a circle of 3/4-inch plywood 2 or 3 inches larger than the flange of the seacock, and bore out the center with your hole saw to form a ring

Dry-Fit

- Shape the bottom of this ring with a rasp or sander to match the inside curvature of the hull around the hole

- The ring serves to reinforce the hull around the hole and to provide a flat surface for the seacock flange

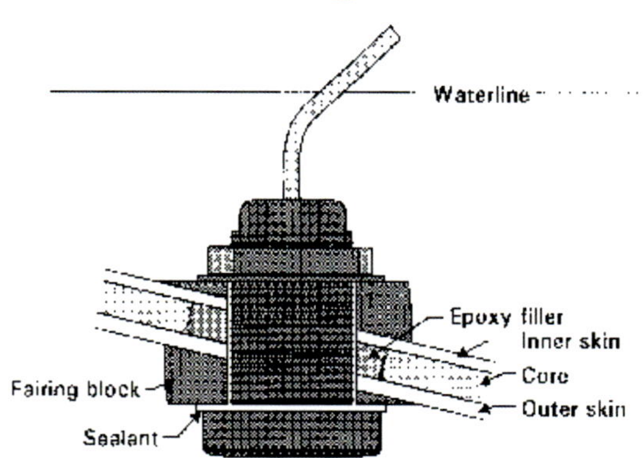

Installation

Once satisfied with the position, drill mounting holes through the hull from inside the boat, using the holes in the mounting flange as a drill guide

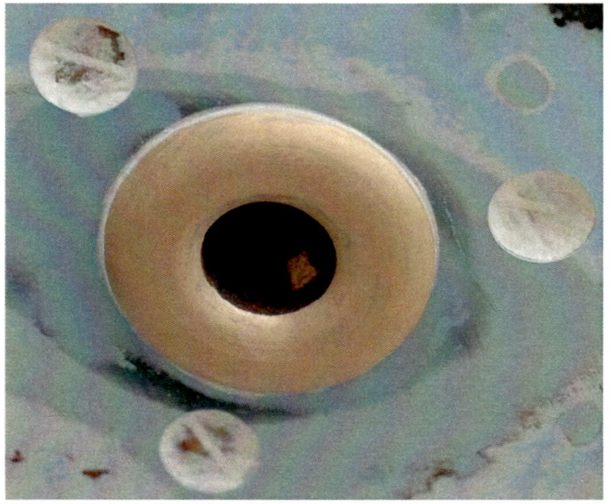

Installation

If the flange is not pre-drilled, drill three evenly spaced holes through the flange and the hull. From outside the hull, countersink the mounting holes

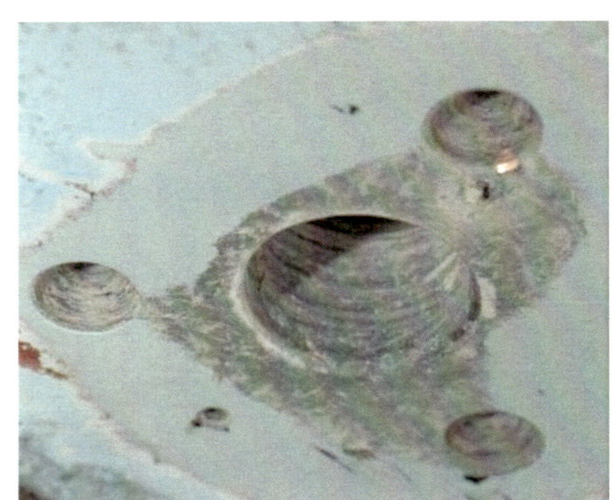

Installation

Unscrew the seacock and remove the through-hull fitting. Give the plywood rings 2 or 3 coats of epoxy (preferred) or varnish before completing the installation

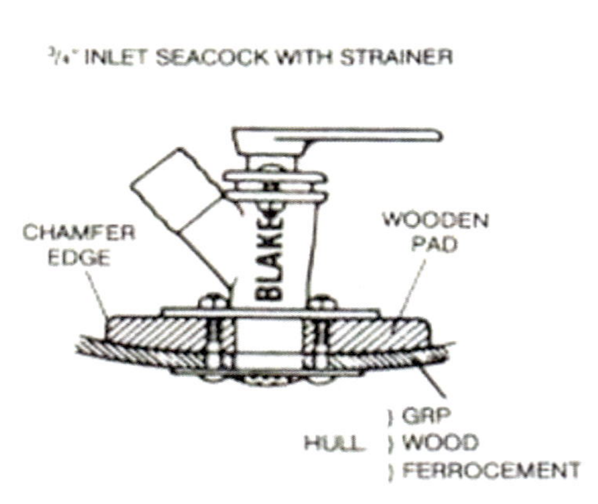

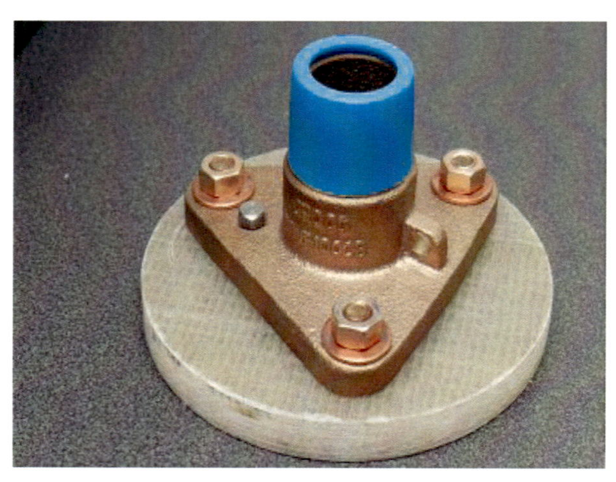

Recommendations

- All piping, tubing or hose lines penetrating the hull below the maximum heeled waterline, shall be equipped with a seacock to stop the admission of water in the event of failure of pipes, tubing or hose
- (Exception 1) Hull penetrations that discharge above the waterline in its static floating position and below the maximum heeled waterline and meet requirements H-27.5.3

Recommendations

Hull penetrations that are not equipped with a seacock (*remember, these may only be above the static waterline*) **shall use reinforced piping or hose that resists kinking or collapse."**

Materials Recommended

- Hardware throughout the raw-water system should be fully compatible with exposure to seawater .

- Usually this means bronze; however, proprietary glass-reinforced plastics such as Marelon are also well suited to seacock and raw-water applications .

- Also, conventional **300-series** stainless steel is *not* ideal because of its propensity for crevice corrosion .

Recommendations

- If you opt to forgo the seacock on any fitting above the static waterline, the hose must be reinforced.

- The seacock must be able to withstand "a 500 pound static force applied for 30 seconds to the inboard end the assembly, without the assembly failing to stop the ingress of water.

- Threads used in seacock installations shall be compatible (e.g., NPT to NPT, NPS to NPS)

Why Is It a Bad Idea ?

This owner simply wanted to remove the gate valve and replace it. When a slight pressure was applied, not even enough to break the threads free, or spin the thru-hull in the hull, the thru-hull snapped off and he was left holding the valve assembly in his hands. WOW!!!!!

Bad Bonding

Hull Strainers

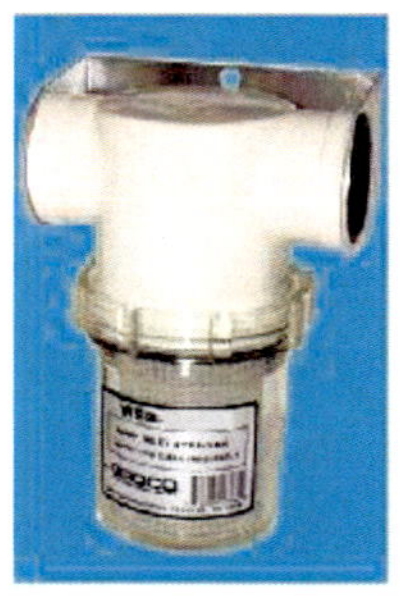

Applications

Raw Water Strainers

- Raw-water strainers are most often associated with the cooling systems of inboard engines

- Every boat that brings aboard outside water to cool the engine, A/C ,Transmission , whether the raw water flows through the engine or through a heat exchanger, needs a strainer to prevent grass and other solids from reaching the pump

- All other raw-water intakes on a boat should also have a filter in the line

Installation

Plumbing a raw water filter is simply a matter of inserting it into the line connected to the seacock

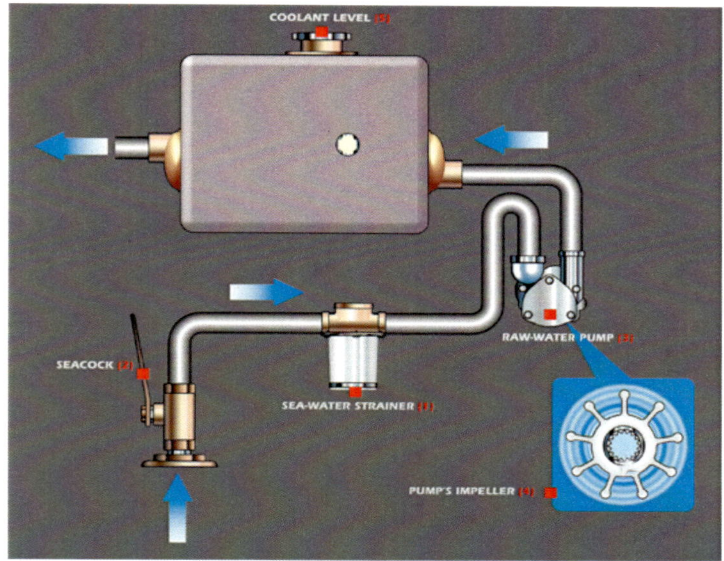

Installation

Bronze filters typically have threaded inlet and discharge ports that you will need to fit with appropriate bronze tailpieces (hose barbs). For leak-free connections, be sure to wrap the tailpiece threads with Teflon plumber's tape before screwing them into the ports

Installation

The mounting location should be as close to the seacock as practical while still allowing for easy servicing of the filter

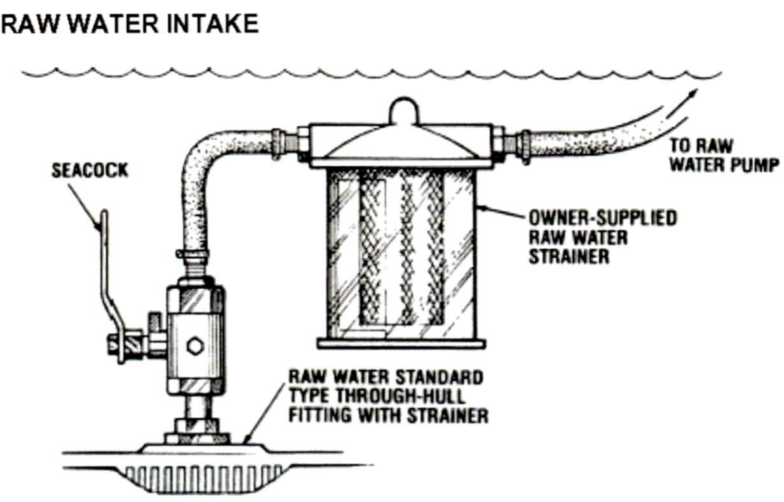

Installation

- The hose connection from the seacock to the filter should be as short and straight as possible. The outlet hose to the pump should likewise not have extra length or bends

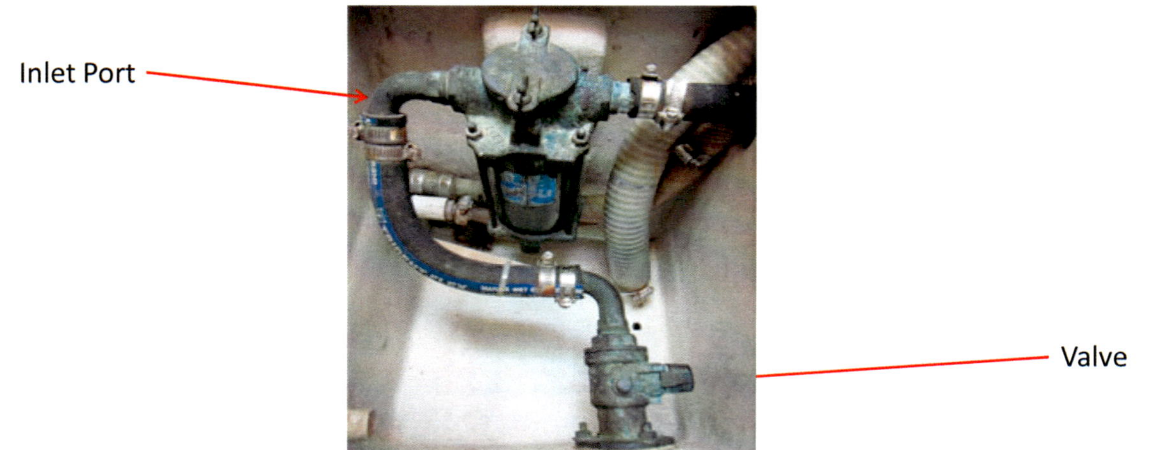

Installation

- Be sure the seacock is connected to the inlet side of the filter-often identified by an arrow pointing on the direction of flow. Use double hose clamps on both the seacock and the filter, even if the filter is above the waterline

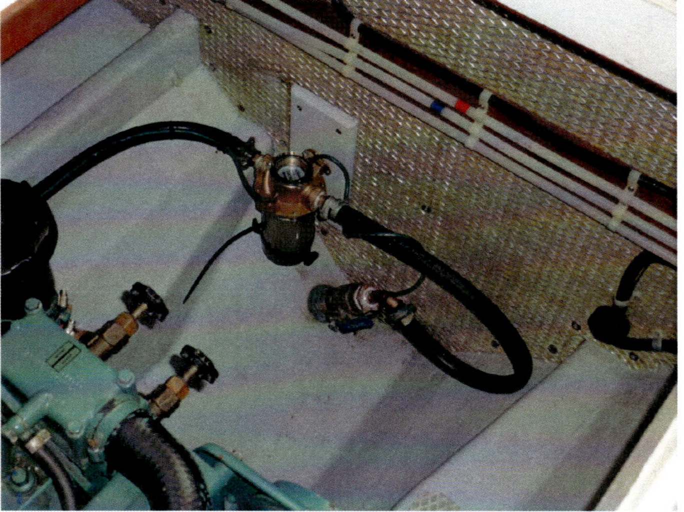

Cleaning

- The easier a strainer is to clean, the more often it will get cleaned
- The preference is for strainers with spin-off caps and lift out baskets. No tools should be required to service the filter

Cleaning

Most modern strainers meet these criteria, but a lot of old strainers must be disassembled to gain access to the screen. If your boat is equipped with such a strainer, you will be doing yourself- and your engine-a favor by replacing it with one that can be cleaned in half a minute

Raw Water Strainers

- Raw-water strainers are most often associated with the cooling systems of inboard engines .

- Every boat that brings aboard outside water to cool the engine , A/C , Transmission, whether the raw water flows through the engine or through a heat exchanger, needs a strainer to prevent grass and other solids from reaching the pump.

- All other raw-water intakes on a boat should also have a filter in the line .

Installation

Bronze filters typically have threaded inlet and discharge ports that you will need to fit with appropriate bronze tailpieces (hose barbs). For leak-free connections, be sure to wrap the tailpiece threads with Teflon plumber's tape before screwing them into the ports

Under Water Lights

- Underwater Lights Types
- Installation
- Wiring

Under Water Lights

- Underwater lighting isn't a necessity on a boat, but it sure does look good when you're moored up at night .

- While cutting a hole in the hull can be a tad scary, several hull lighting options are available that fasten on the outside of the hull, in this case quite conveniently on trim tabs

Under Water Lights Installation

Select the areas on your transom to install each side light. If your boat has trim tabs, install the LED lights directly over the inside corner of each trim tab. The best location for your middle light (if desired) depends on what type of engine situation you have

Under Water Lights Installation

- The location you choose must have access to the inside of the boat on the other side of the transom

- Lay the bracket against the outside of the transom and mark the screw holes with a marker. Drill a hole as a pilot to make sure no wires, stringers, or equipment block the desired area

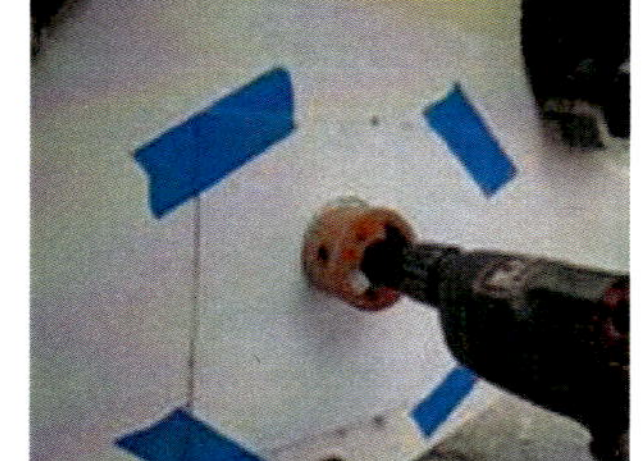

Under Water Lights Installation

Use a marine grade underwater sealant to coat the screw openings on the light plate and apply a generous amount of sealant to the wire opening. Put the plate on the hull and screw on tight

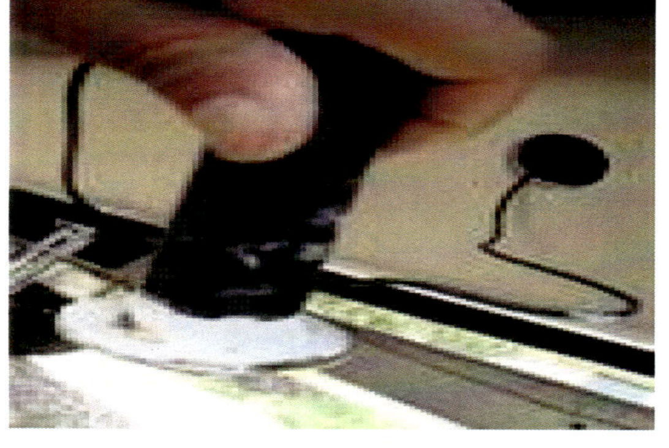

Under Water Lights

Repeat the light application for all required units. Run the wires up the side of the boat. Secure loose wires with plastic ties to keep them with other electronics. Connect wiring to the navigation light switch at the helm, or any desired toggle switch

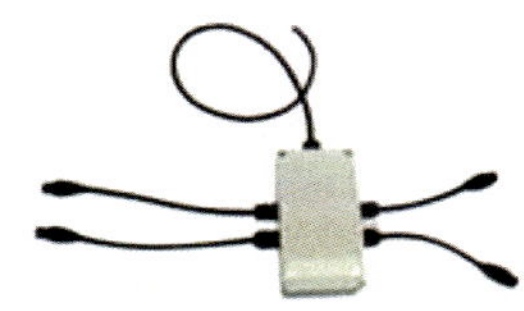

Series or Parallel Wiring

- The rule of thumb once again is , ! Read the instructions!. Verify the amps consumed per each light and the wire size recommended
- For direct current systems I recommend install a common positive bus bar at the transom boat and then each light can be powered from the bar individually
- In this way each element will work independently and protected for a separate fuse

Chapter 2
Hydraulic Systems

- Marine Hydraulic System
- Hydraulic Components
- Hydraulic Circuits
- Principle of Hydraulic Drive
- Hydraulic or Pneumatic
- Pascal Law
- Hoses ,Pipes and Lines
- Hydraulic Pumps
- Pressure, Flow , Speed and Efficiency
- Hydraulic Pumps Calculation
- Solenoid Valves
- Hydraulic Cylinders
- Hydraulic Fluids , Symbols and Diagrams

Hydraulic Systems

- Hydraulic systems have the ability to multiply torque or apply force in a simple way

- Mechanical systems would require an intricate system of gears, chains, pulleys and levers, to move machinery at a distance from the engine

- Hydraulic systems, however, can transmit force from a force engine to the place where it needs to be in order to do the work simply by stringing hydraulic hoses between the two

Hydraulic Fluid , Pipes & Hoses

- Fluids transmit force effectively because they do not compress
- The force that is applied at one end of a hydraulic hose travels to the opposite end of the hose with little loss of power. Changes in size of hoses along the way can increase or decrease the force applied at the opposite end

What is a Hydraulic Circuit

Is a close circuit of mechanisms operated by the pressure transmitted for a fluid

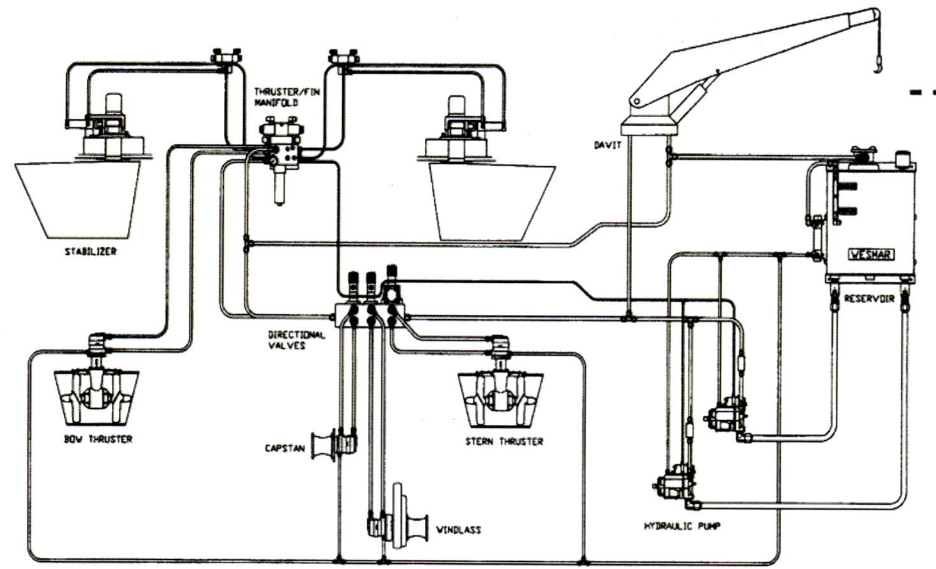

Hydraulics Today

More and more, we're seeing hydraulics in the trawler lifestyle. Cruising generators, active fin stabilizer systems, windlasses, dinghy lifts, bow and stern thrusters, get home drives, steering systems, main propulsion systems-all play an active part in making our boats safer, more comfortable, and easier to operate

Why Hydraulic??

- Because it can't be compressed, a confined fluid is incredibly strong, can be minutely adjusted in any direction, yet is still easily controlled.

- Most marine applications use small, easily installed components, with hydraulic lines run through the boat. Properly engineered, hydraulic systems can be stalled without damage, are reversible, and offer a lot of usable power when integrated into a multi-function system

Hydraulic Components

Generally speaking, all hydraulic systems have the same system elements, whether they are used for windlass, fin stabilizer system, get home, steering, or thruster applications. They include pump, tank and filter, lines, controls , heat exchanger and output device

Pneumatic Circuits

- Pneumatics is study of mechanical motion caused by pressurized gases and how this motion can be used to perform engineering tasks

- In order to affect mechanical motion, pneumatics employs compression of gases, based on the working principles of fluid dynamics in the concept of pressure

- A pneumatic circuit consisting of active components such as gas compressor, transition lines, air tanks, hoses, open atmosphere, and passive components

A Basic Hydraulic Circuit

Any hydraulic circuit is a closed system where the hydraulic fluid under pressure, after performing the work, returns back to the same system

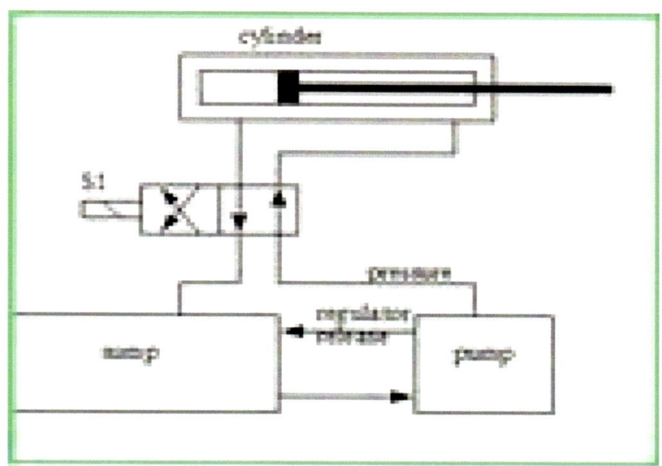

A Basic Hydraulic Circuit

To obtain any hydraulic fluid under pressure, we require a pump unit with necessary pipelines and valves. It is highly essential to have an accumulator on the pump's discharge side

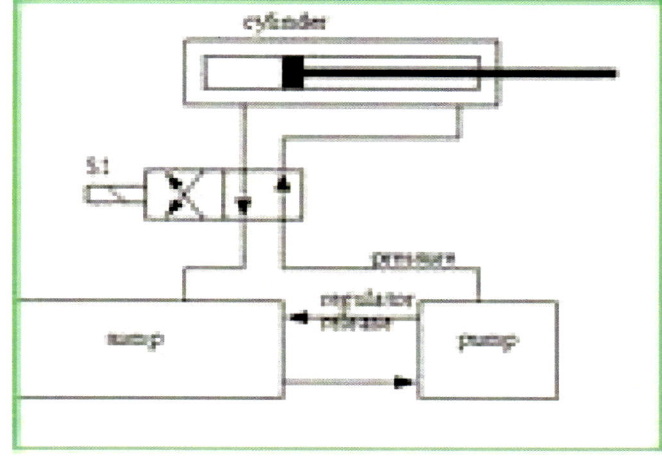

Typical Boat Hydraulic System

- Steering Systems
- Bow and Stern Thrusters
- Davits and Cranes
- Anchor
- Stabilizers
- Garage Doors
- Swim Ladders and Platforms
- Gangways and passerelles

Principle of a hydraulic drive

- The popularity of hydraulic machinery is due to the very large amount of power that can be transferred through small tubes and flexible hoses, and the high power density and wide array of actuators that can make use of this power.

- Hydraulic machinery is operated by the use of hydraulics, where a liquid is the powering medium.

Principle of a hydraulic drive

- Pascal's law is the basis of hydraulic drive systems. As the pressure in the system is the same, the force that the fluid gives to the surroundings is therefore equal to pressure x area. In such a way, a small piston feels a small force and a large piston feels a large force. **P=F/A**
- $P_1 = P_2$ $\qquad$ $F_1/A_1 = F_2/A_2$
- The Power that is applied at one point is transmitted to another point using an incompressible fluid.

Hydraulic or Pneumatic ?

Hydraulic systems are used to get high output pressures but with a low output speeds , instead that pneumatic systems are used to get low output pressures with high speeds

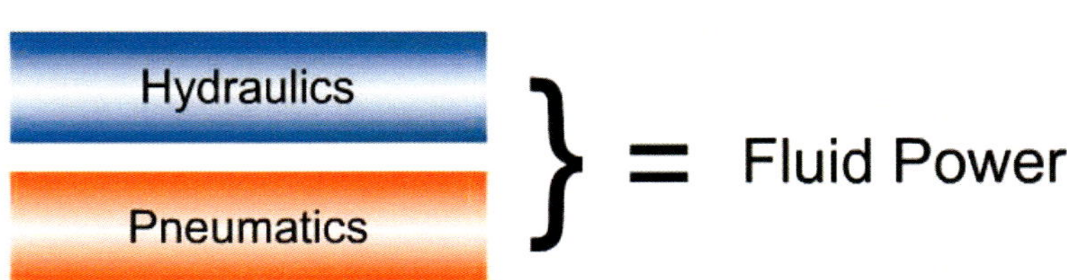

Pascal Law

Pascal's law states that when there is an increase in pressure at any point in a confined fluid, there is an equal increase at every other point in the container.

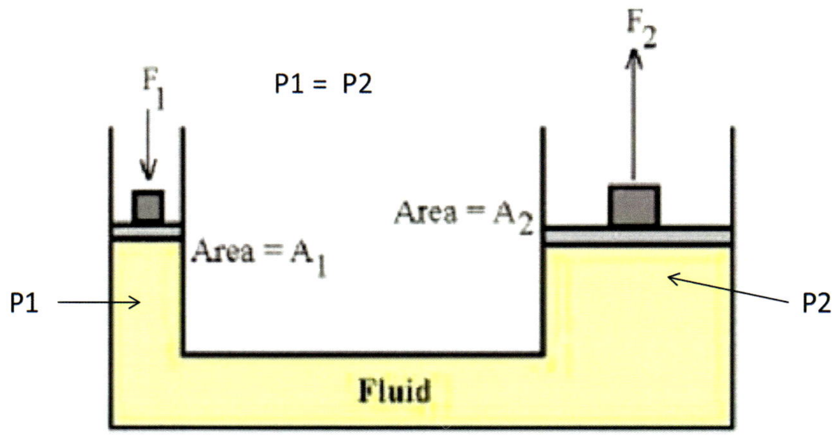

Pascal Law

The cylinder on the left shows a cross-section area of 1 square inch, while the cylinder on the right shows a cross-section area of 10 square inches

Remember that the pressure is constant in both sides of the circuit $P1 = F1/A1$ $P2 = F2/A2$

$$F1/A1 = F2/A2$$

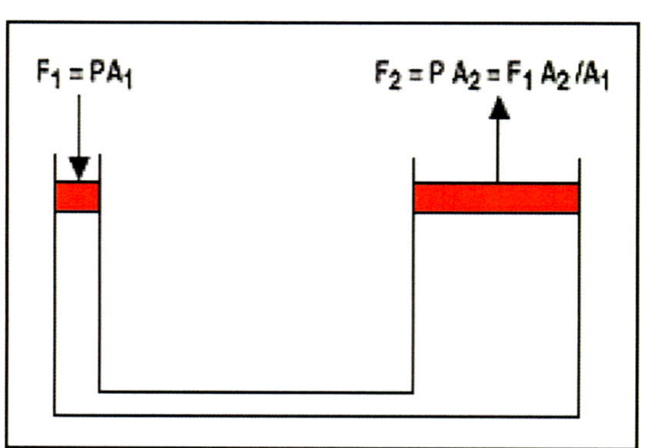

Pascal Law

How much should be F_2 if F_1 is 1Lb , the A_1 is 1 Square in and the A_2 is 10 Square in?

Remember that the pressure is constant in both sides of the circuit $P_1 = F_1/A_1$ $P_2 = F_2/A_2$

$$F_1/A_1 = F_2/A_2$$

$$1 \text{ Lb}/ 1 \text{ in} = F_2/ 10 \text{ in}$$

$$\text{Then } F_2 = 10 \text{ Lb}$$

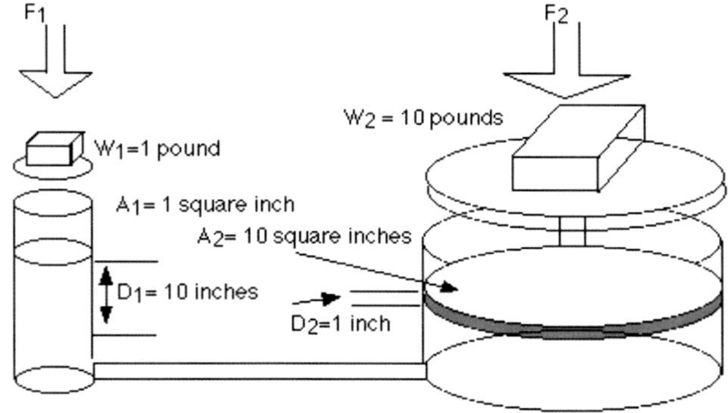

Pressure and Work

As a consequence that the pressure is constant. The **work** done at the cylinder number one should be the same than the work done in the cylinder two

$$P_1 = F_1/A_1 \qquad P_2 = F_2/A_2$$

$$W_1 = F_1 \times D_1 \qquad W_2 = F_2 \times D_2$$

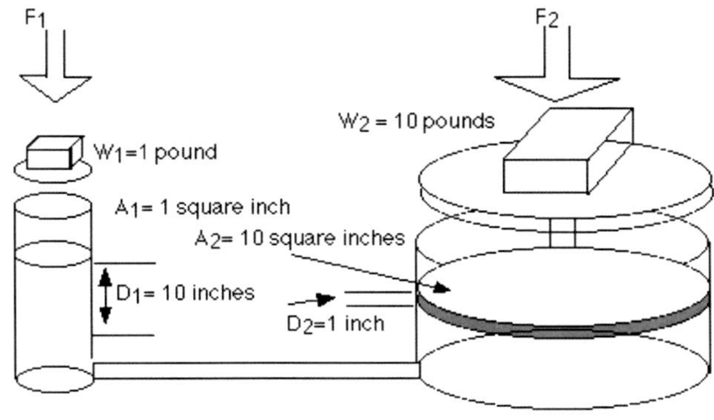

Pressure & Work

How much should be distance traveled by the second piston (D2). if F1 is 1 Lb and the distance traveled by the first piston is 10 inches and the force exerted by the second piston is 10 pounds?

Remember that the work is constant in both sides of the circuit $W_1 = F_1 \times D_1$ $W_2 = F_2 \times D_2$

$$W_1 = 1 \text{ Lb} \times 10_{in} = 10 \text{ Lb / in}$$

$$W_1 = W_2 = 10 = F2 \times D2$$

Then $D_2 = 10 \text{ Lb} / 10 \text{ Lb / in}$

$$D_2 = 1 \text{ in}$$

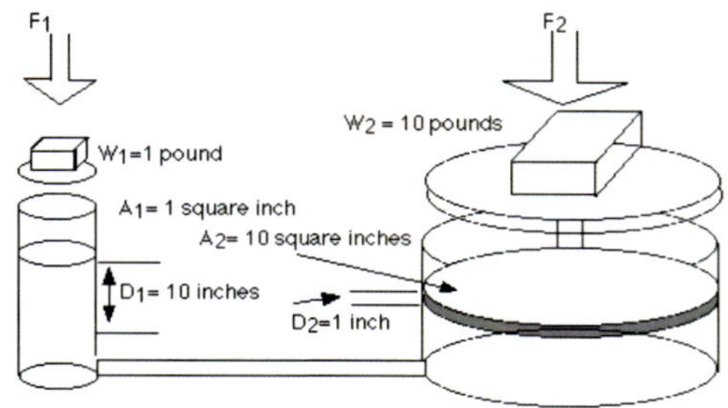

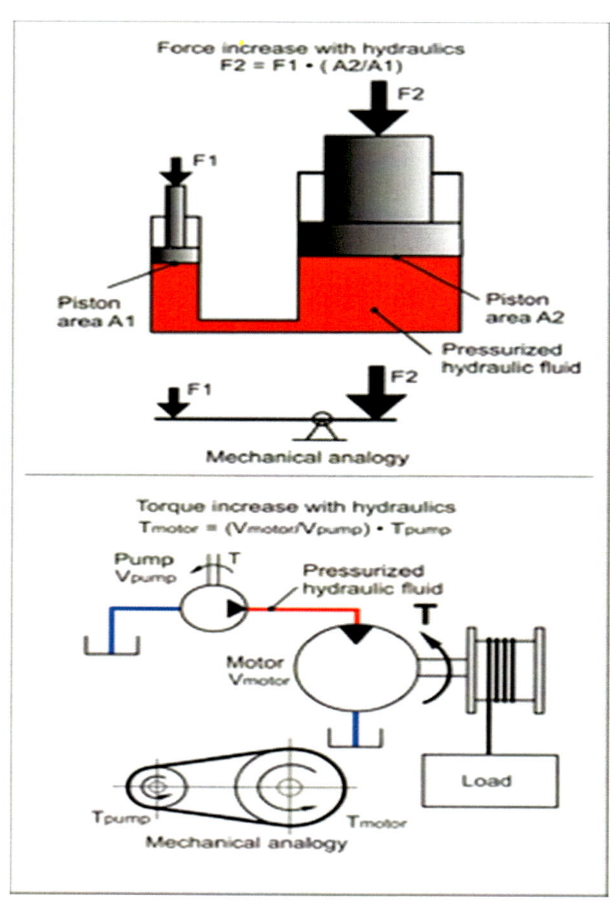

Pascal Law Exercise

- A hydraulic press has an input cylinder 1 inch in diameter and an output cylinder 6 inches in diameter.
  - Assuming 100% efficiency, find the force exerted by the output piston when a force of 10 pounds is applied to the input piston
- If the input piston is moved through 4 inches, how far is the output piston moved?.
- Answers: a) 360 Lb   b)1/9 in.

Pascal Law Exercise

- A car has a weight of 2500 pounds and rests on four tires, each having a surface area of contact with the ground of 14 square inches. What is the pressure the ground experiences beneath the tires that is due to the car?

- Answer: a) 44.64 PSI

Pascal Law Exercise

- What pressure does a 130 pound woman exert on the floor when she balances on one of her heels? Her heels have an average radius of 0.5 inch .

- Answer = 165.52 PSI .

Hydraulic Jack

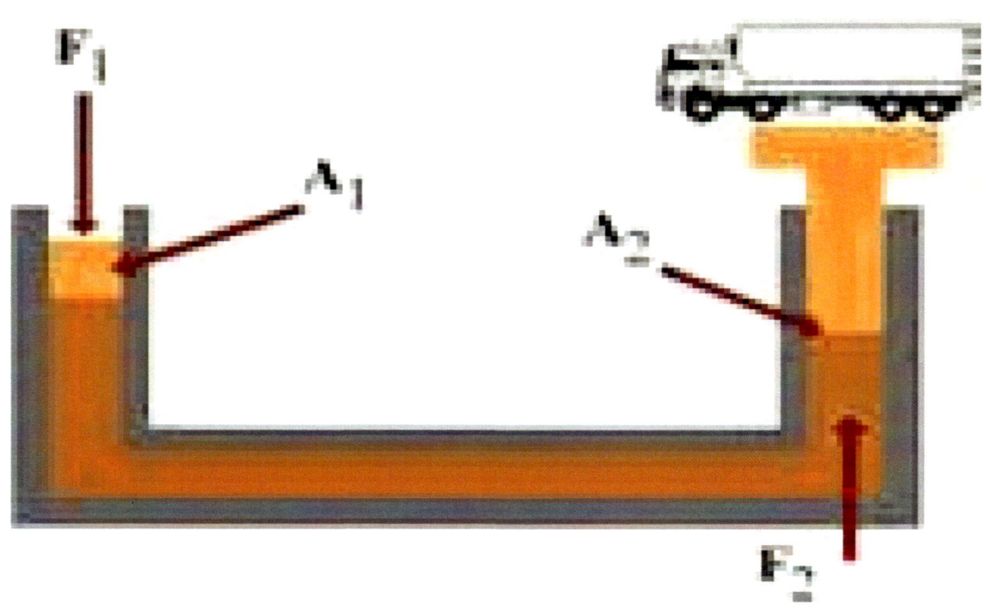

Hydraulic brake

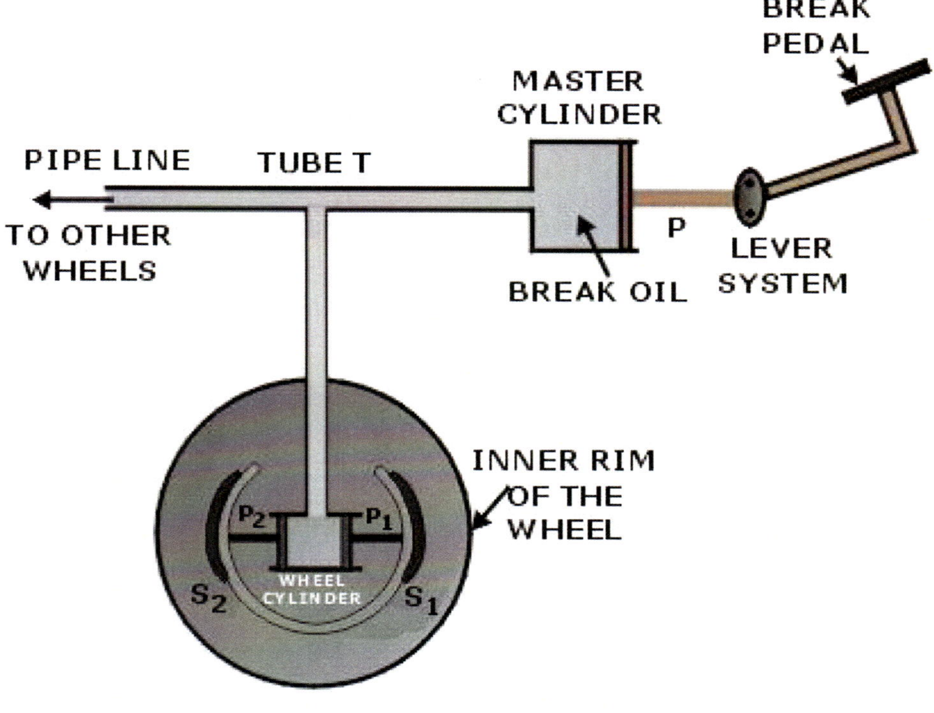

Hydraulic Brake

Hydraulic Press

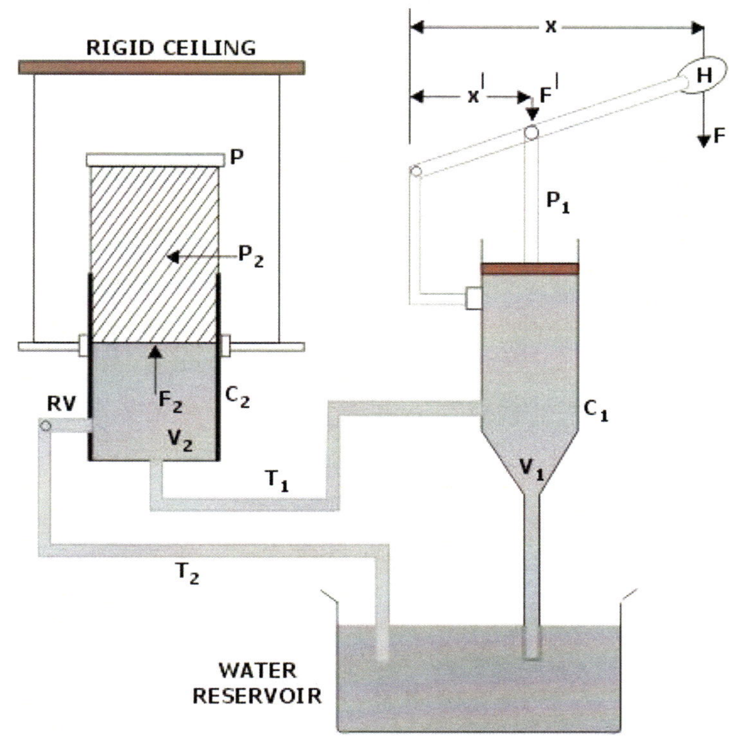

Hydraulic Press

The Pump and Reservoir

- The pump used is usually a positive displacement type, especially a **Gear Pump**
- The pump takes suction from the reservoir, which is filled with the hydraulic oil. The reservoir has a level gauge, from which the oil level can be witnessed

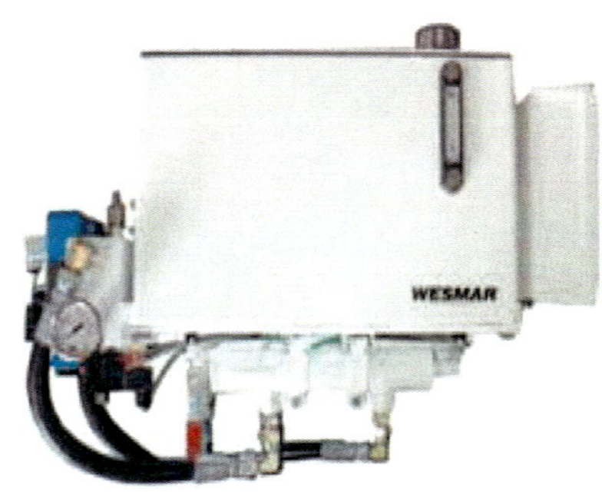

The Relief Valve

The relief valve, relieve excess oil pressure back into the reservoir tank, thus maintaining safety of the system. There are some systems which have two relief valve arrangements, one draining into the reservoir and other an external relief valve

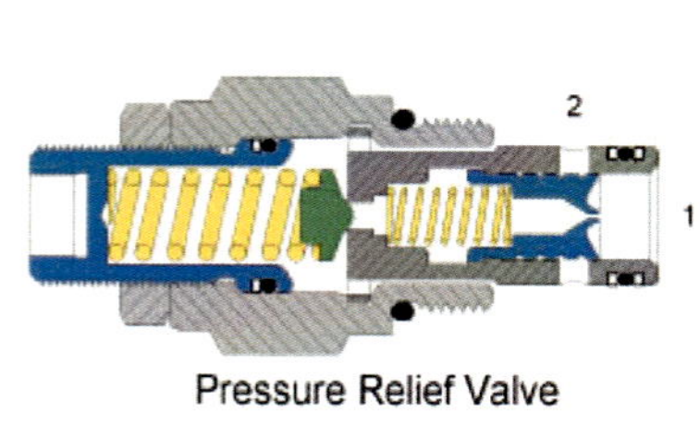

Pressure Relief Valve

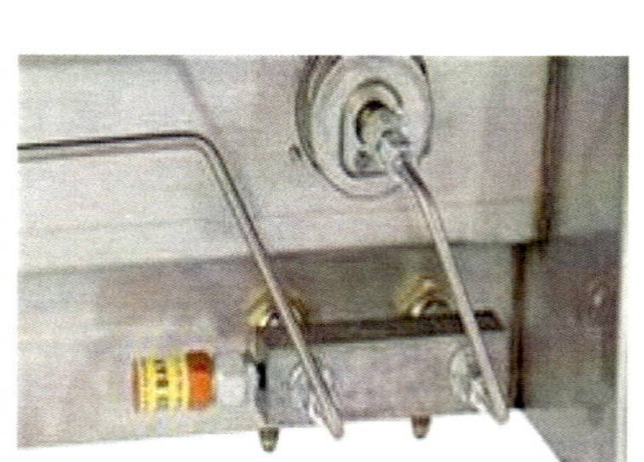

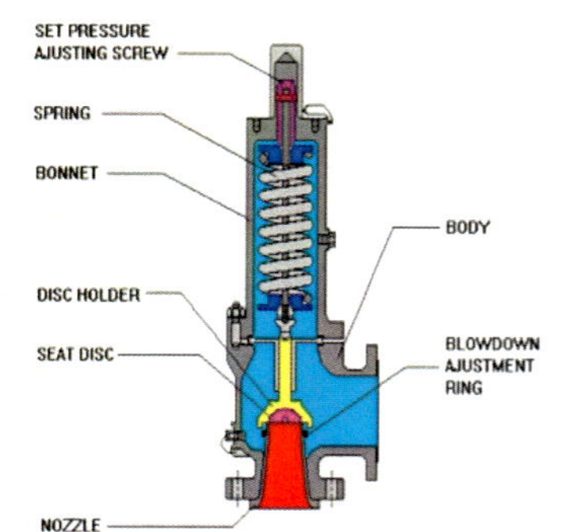

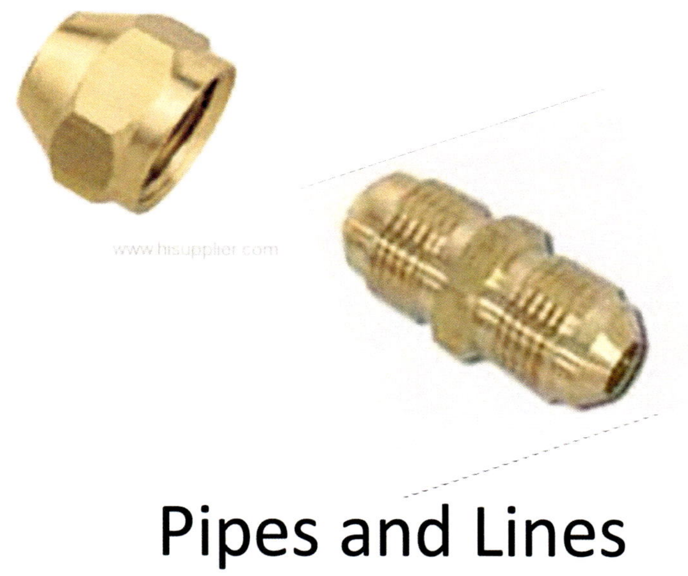

Pipes and Lines

Flare Fittings

The Hydraulic Hoses

A hydraulic hose is a high-pressure, synthetic rubber, thermoplastic or Teflon reinforced hose that carries fluid to transmit force within hydraulic machinery

Hose Ratings

Every hose has specific ratings for specific types of fluid they are designed to carry, working temperature ranges and pressure limits for that specific hose

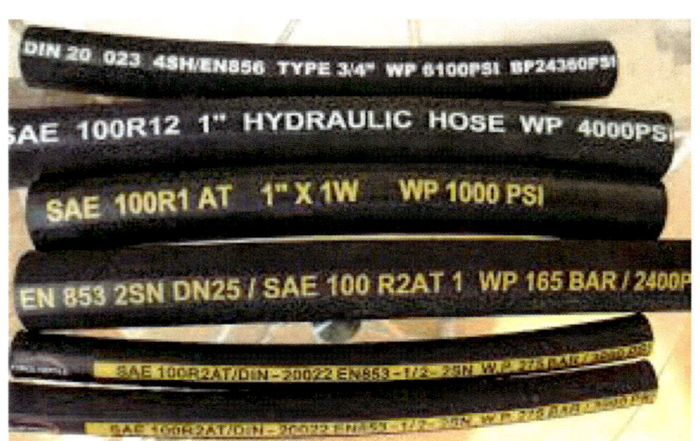

Hose Ratings

Usually they are printed on the hose or fittings. In some case they print a model number on the hose and provide a spec sheet for the various models

Caution

Hydraulic systems operate under high pressure to drive machinery. Hoses that fail at high pressures can whip about with extreme violence and injure bystanders or machine operators. Hydraulic hoses should be checked and replaced according to manufacturers recommendations

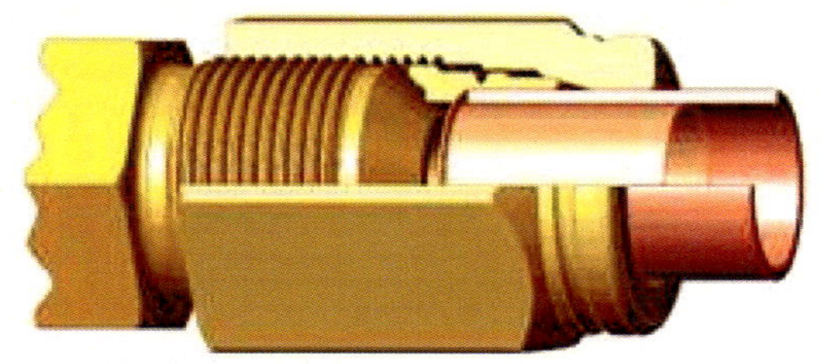

Flare Nut & Fitting

Flare Nuts

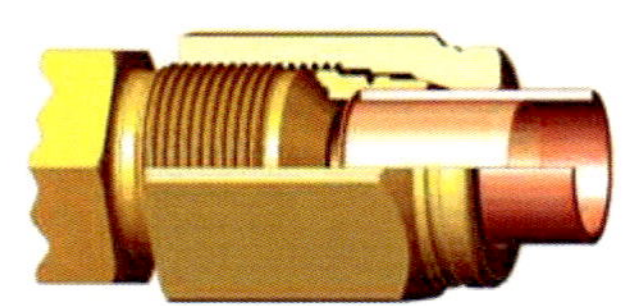

Flare Fittings

Flare Configurations

Flare Tool

 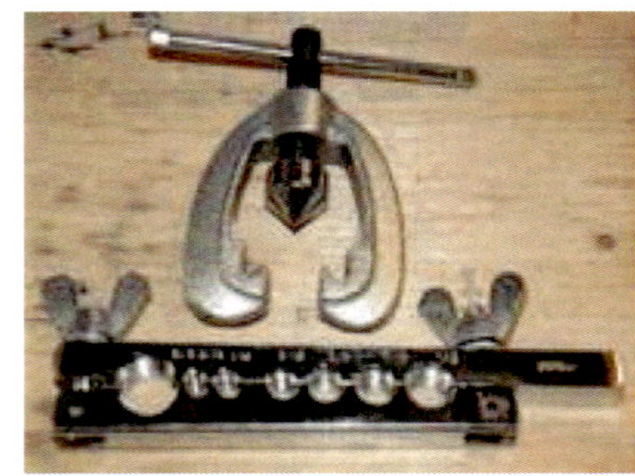

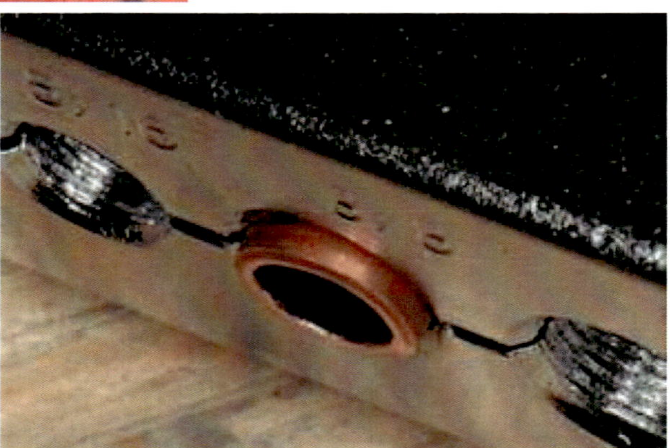

Soldering Pipes

Hydraulic Pumps

- Pumps can be classified as positive displacement pump and non-positive displacement pump.
- The hydraulic system frequently uses positive displacement pump. The pressure used in a positive displacement pump can range up to 5000 psi.
- These pumps function by means of forcing a volume of fluid from the pump's inlet pressure into the pump's discharge zone.

Types of pumps

- There are three basic types of pumps found in hydraulic systems. The gear pump, Vane or Rotary Pump and Piston pump .
- The flow of hydraulic fluid through a pump is measured in gallons per minute, or GPM. This is useful information, and is used in calculating a pump's input horsepower requirement. That formula is:
- *Horsepower (hp) = Pressure (psi) x GPM / 1714*
- For example, a fluid pressure of 2,500 psi in a pump with a 30 GPM rated flow requires 43.75 hp to create this pressure and flow

Hydraulic Pumps

- Transfer Pump

- Trim Tabs Pump

- Helm steering pump

Hydraulic Pumps

- Auto pilot Pump

- Mercruiser Trim Pump

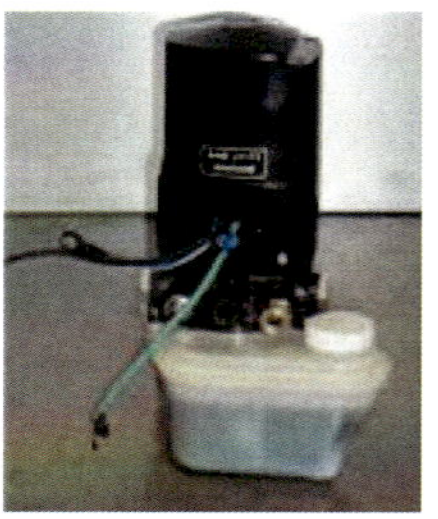

- Helm Pump

Hydraulic Pumps

- Engine steering pump

- Accessories Pump

- Fresh Water Pump

Hydraulic Pumps

- Macerator Pump

How Hydraulic Pumps Work

- Generally, hydraulic pumps work in two functions by using a mechanical action.

- First, its mechanical action will create a vacuum at the pump inlet which will allow atmospheric pressure to force the liquid from the reservoir into the inlet line to the pump.

- Second, its mechanical action will deliver this liquid to the pump outlet and will force it to the hydraulic system.

Gear Pumps

- Gear pumps (with external teeth) (fixed displacement) are simple and economical pumps.

- These pumps create pressure through the meshing of the gear teeth, which forces fluid around the gears to pressurize the outlet side

- This pump is suitable for pressures below 3000psi

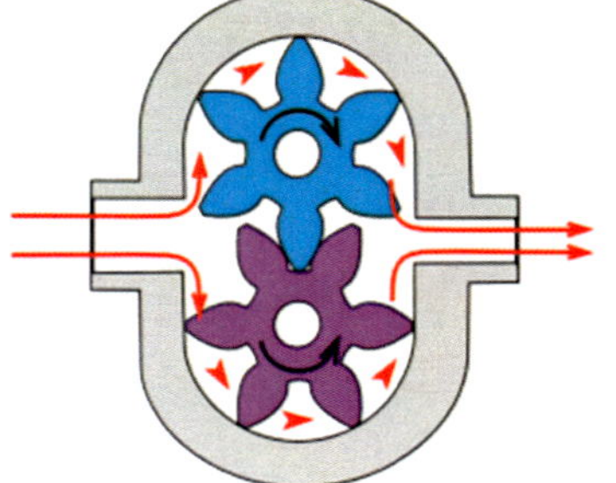

Rotary vane pumps

- Also called <u>Fixed displacement pumps</u> have higher efficiencies than gear pumps, but are also used for mid pressures up to 180 bars in general

- These vane pumps are in general constant pressure or constant power pump

- Typically, maximum pressure for external gear and vane pumps are from 2,000 to 4,000 psi

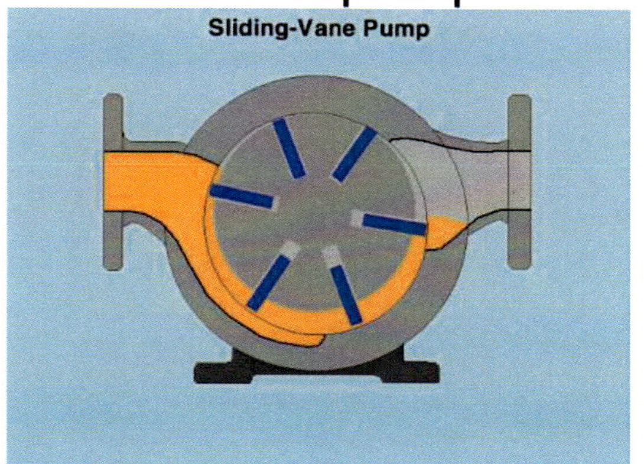

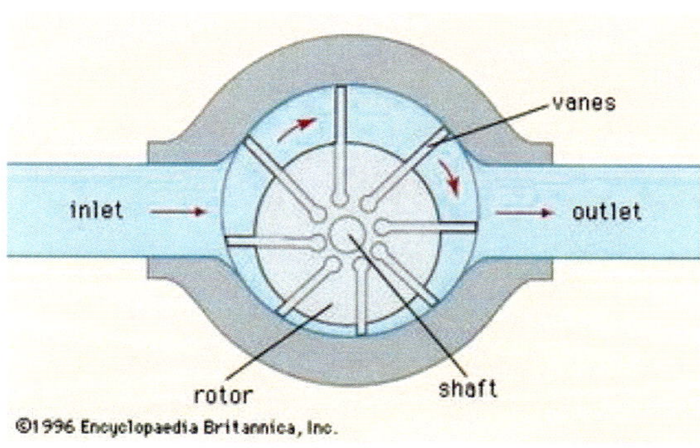

Screw pumps

- They are a double <u>Archimedes' screw</u>, but closed. This means that two screws are used in one body.

- They were used on board ships where the constant pressure hydraulic system was going through the whole ship, especially for the control of <u>ball valves</u>, but also for the steering gear and help drive systems

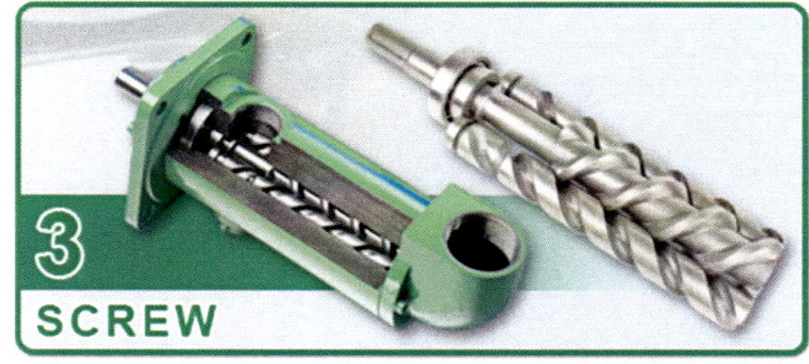

PRESSURE

- As a pump rotates, it develops a partial vacuum on the inlet ("suction") side, permitting fluid under atmospheric pressure in the reservoir to flow into the pump inlet. Then the pump ejects this fluid, usually at a <u>pressure higher than atmospheric.</u>

- It is worth noting that a pump does not create pressure. It merely moves fluid, causing the flow.

Figure 4-7.—Lobe pump.

PRESSURE

- Pressure is created by the load on the fluid; if no load exists, the fluid has very little pressure.

- As the load is placed on the fluid, the pressure at the outlet side of the pump increases to a value that is normally indicated as the pump maximum. Therefore, a 3,000-psi pump is a unit that can maintain flow against a load of 3,000 psi.

- Pump pressure rating is generally limited by the capability of the pump to withstand pressure without undesirable increase in internal leakage.

FLOW

- The second most important consideration in selecting a pump is its size and delivery.

- Size is usually expressed as volumetric flow output (gpm). Other words with the same meaning are flow, size, capacity, or delivery rate.

- pumps used in mobile applications are generally tested at 1,200 rpm, at an outlet pressure of 100 psi and atmospheric inlet pressure.

SPEED

- A third consideration is the speed rating, which may be limited by the ability of the pump to fill without cavitating or by other mechanical considerations.

- The permissible speed range and inlet pressure requirements for any design are usually clearly defined.

- The speed of the fluid is usually expressed in m/sec

What is volumetric efficiency?

- It is a measure of a hydraulic pump's volumetric losses through internal leakage.
- It is calculated by dividing the pump's actual output in liters or gallons per minute by its theoretical output, expressed as a percentage.
- For example, a hydraulic pump with a theoretical output of 100 GPM, and an actual output of 94 GPM is said to have a volumetric efficiency of 94%

Hydraulic Pumps Calculation Formulas

- **[Q] Flow = (m^3/s) .**
- **Q = n * Vstroke *η vol**
  - Q= Flow in m^3/s
  - n = revs per second = (RPM/60)
  - Vstroke = swept volume in m3
  - η = volumetric efficiency
- **Q = n * Vstroke *η vol = v * A.**
  - v= Speed of the fluid in m^3/s
  - A= Output Pipe Area

<u>Solenoids</u>

Marine Applications

Solenoid Valve

- A **solenoid valve** is an electromechanical valve for use with liquid or gas
- The valve is controlled by an electric current through a <u>solenoid</u> coil.

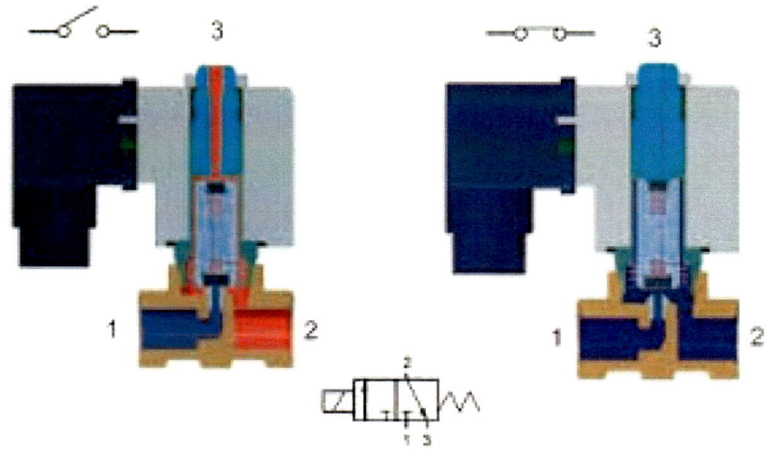

Solenoid Valves

Solenoid valves are the most frequently used control elements in fluidics. Their tasks are to shut off, release, dose, distribute or mix fluids

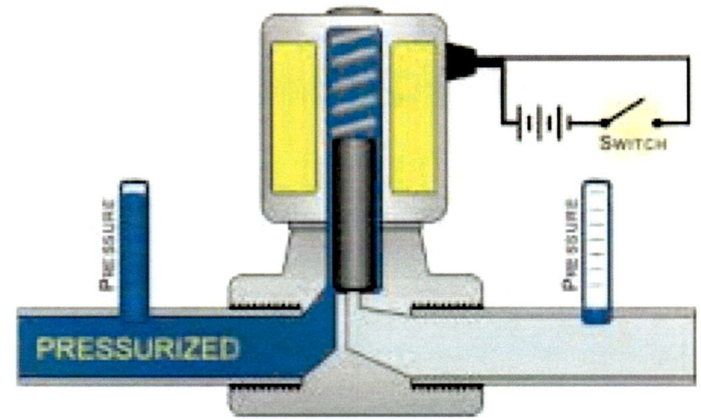

Common uses

Solenoid valves are used in fluid power pneumatic and hydraulic systems, to control cylinders, fluid power motors or larger Boat Applications

Marine applications

- They works with AC or DC power and voltage from 6 Volts until 220 Volts . In marine applications they are used to control :
  - Input fuel to the Fuel Injection Pump (Diesel Engines)
  - Input fuel to the Common Rail Fuel System (Gas & Diesel)
  - Charge and Discharge in Toilet Systems
  - Forward and reverse control in engine transmissions
  - Sense of rotation in Power steering, Stabilizers , Windlass and hydraulic Thruster systems
  - Up and Down control in Trim Tabs systems

Solenoid valves

Servo Valve (Solenoid)

2 Way Solenoid Valve

Two-way valves are shut-off valves with one inlet port and one outlet port. In the de-energized condition, the core spring, assisted by the fluid pressure, holds the valve seal on the valve seat to shut off the flow

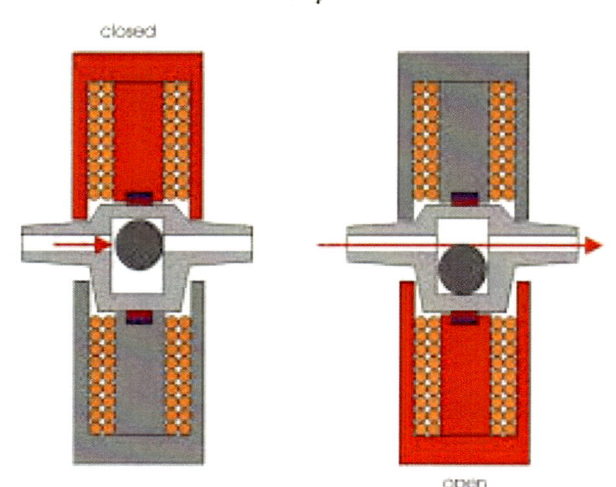

Solenoid valves

They are powered from : Ignition switch , momentary push buttons, control box relays , toggle switches or from electronic control units (ECU)

Hydraulic Cylinders

Is an equipment designed to convert the potential energy accumulated in a fluid into kinetic energy. They are classified according to the fluid used to transmit the power, by the output torque produced , by the time of response to the input signal and by the length of the extended rod

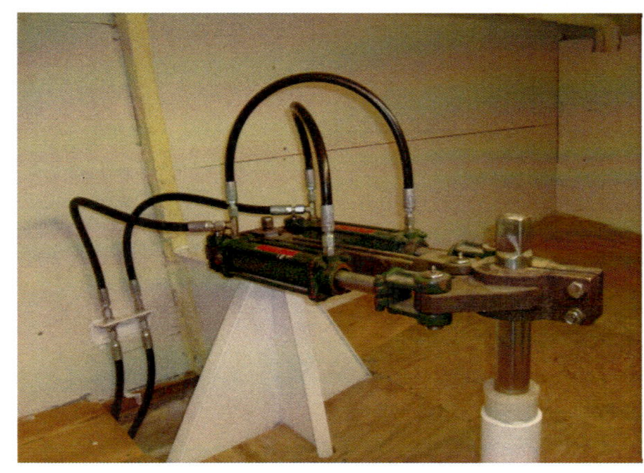

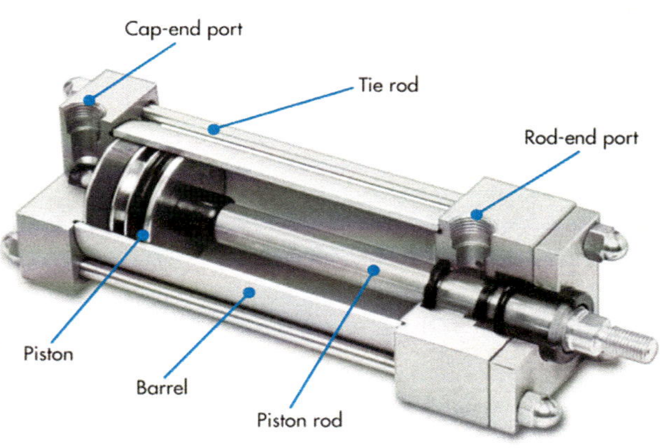

The Output Torque

According with Pascal law if you have a constant pressure in the system due to the pump capacity ,just varying the diameter at the piston then the output force at the piston rod will be modified and consequently the output torque can be altered . P1 = P2

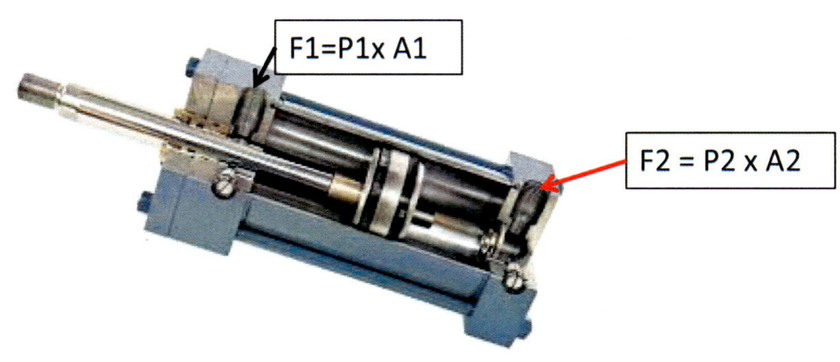

The Output Torque

In other words for the same input fluid pressure you can select different actuators that produce different output forces

Cylinder and Pump

The **mounted style** integrated the pump and the motor in a compact unit , while the **remote style** connect the pump and the motor by means of flexible lines , however the efficiency is the same

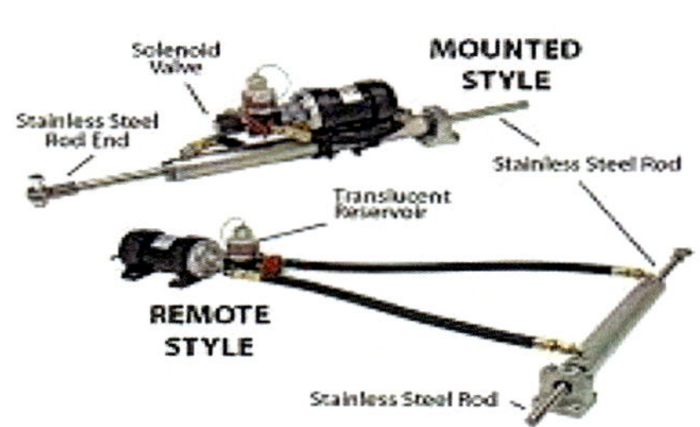

Single or Double acting cylinder

Depending if the cylinder is able to receive fluid in both fluid ports, the actuator is called double action cylinder . otherwise if the cylinder only receive fluid from one side and return by means of a spring the actuator is called single acting spring return cylinder

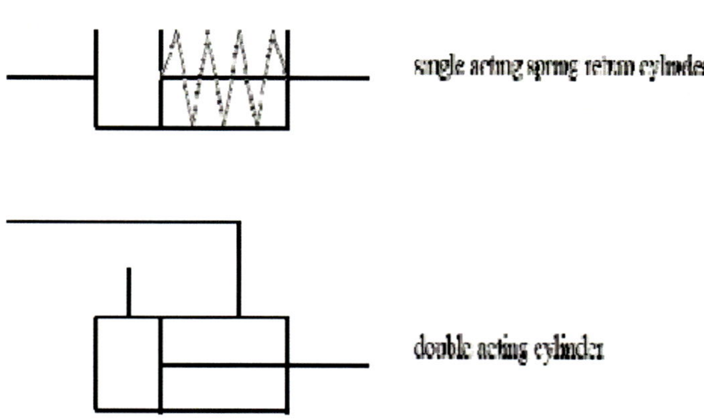

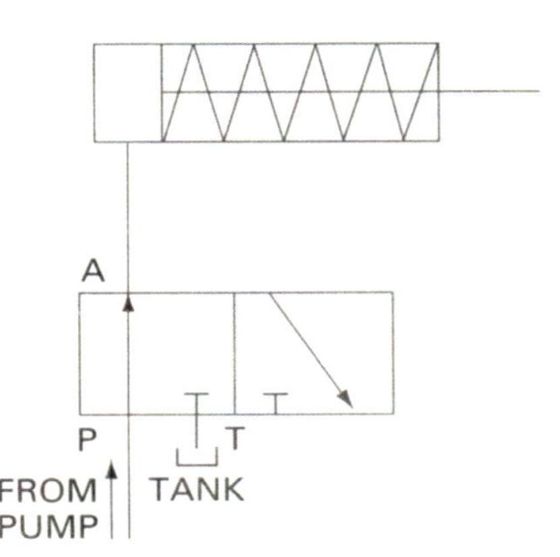

Double Acting – Single Shaft

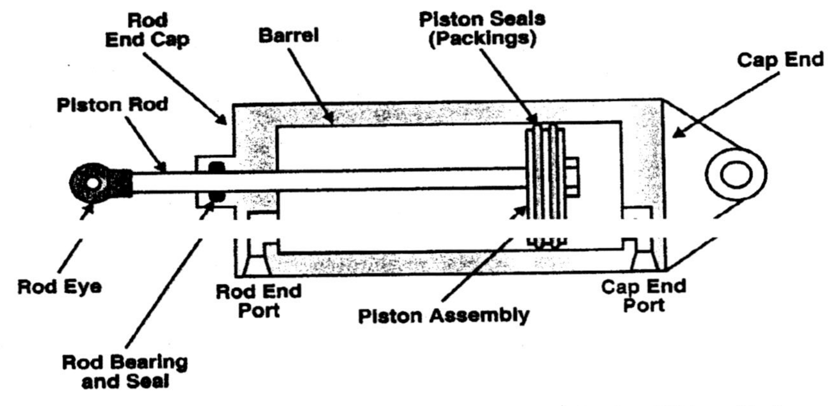

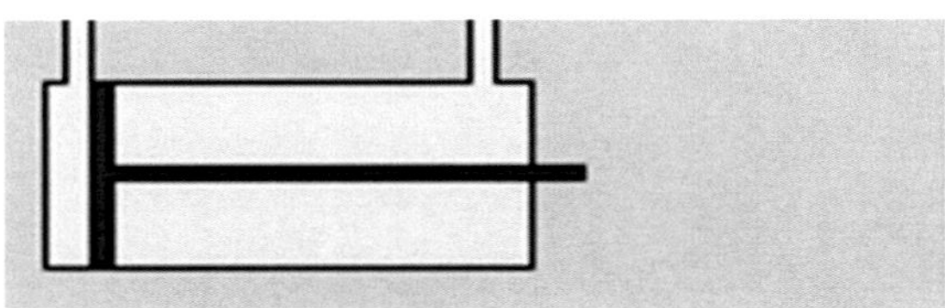

Double Acting / Double Shaft

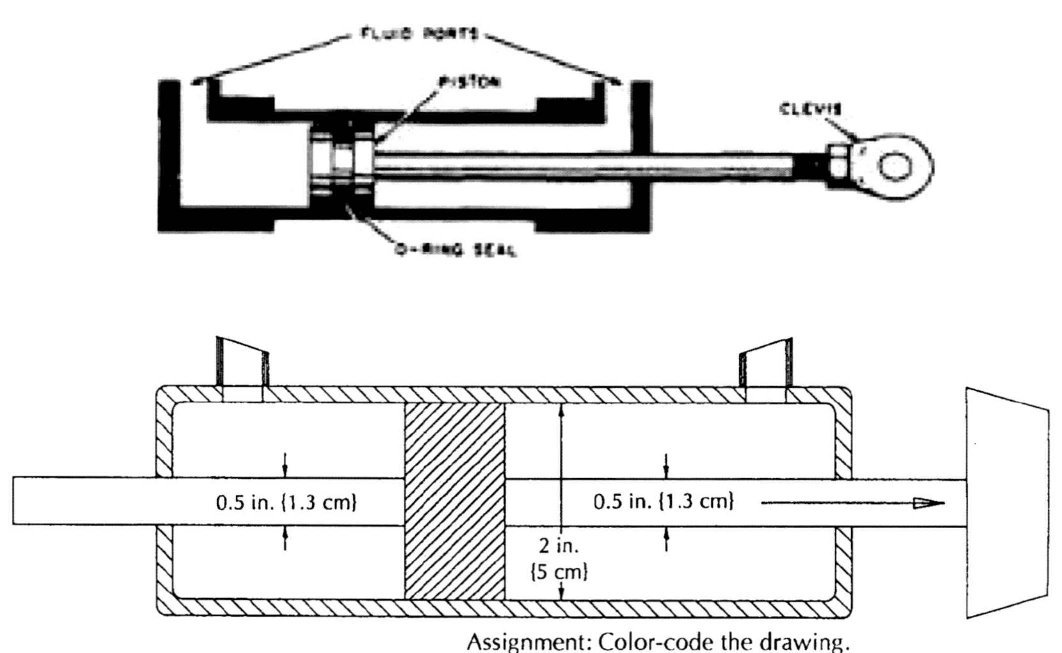

Assignment: Color-code the drawing.

Hydraulic Cylinder Symbols

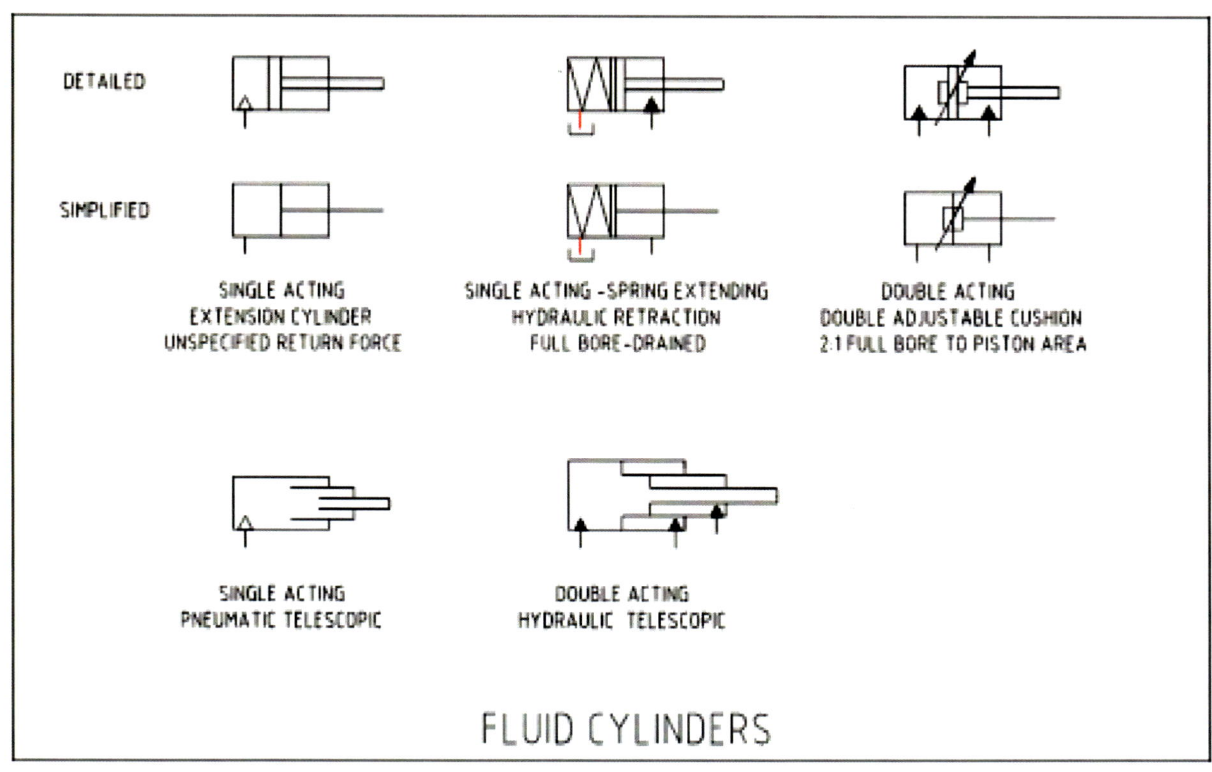

Filling and Bleeding the System

Verify that all bleeder valves on the cylinders , helm pumps , body valves and clutch slaves are closed, and linkages disconnected

Filling and Bleeding the System

Remove fill port plug from the reservoir and fill the reservoir within one inch of the top of the sight tube with hydraulic fluid. Replace fill port plug

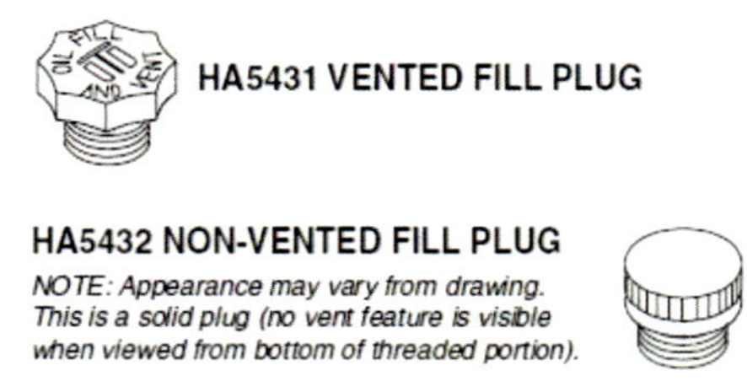

Filling and Bleeding the System

Remove fill port plug from the reservoir and fill the reservoir within one inch of the top of the sight tube with hydraulic fluid. Replace fill port plug

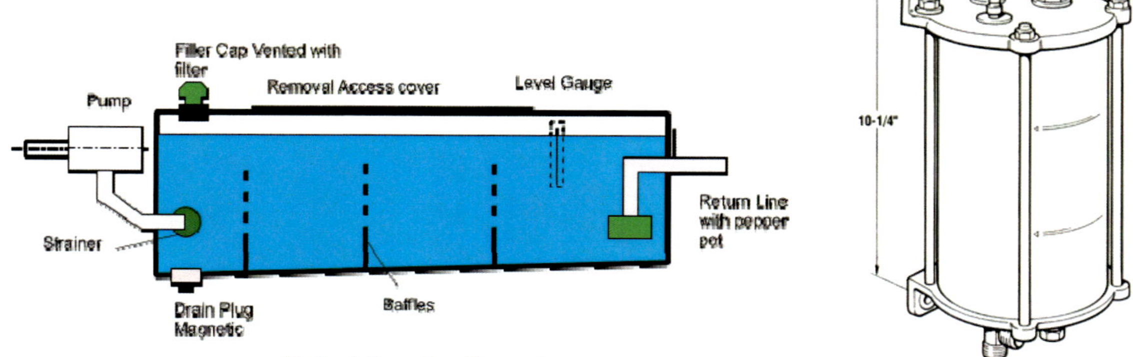

Filling and Bleeding the System

Pressurize reservoir according with the manufacturer recommendations through the air filter valve in the top of the reservoir

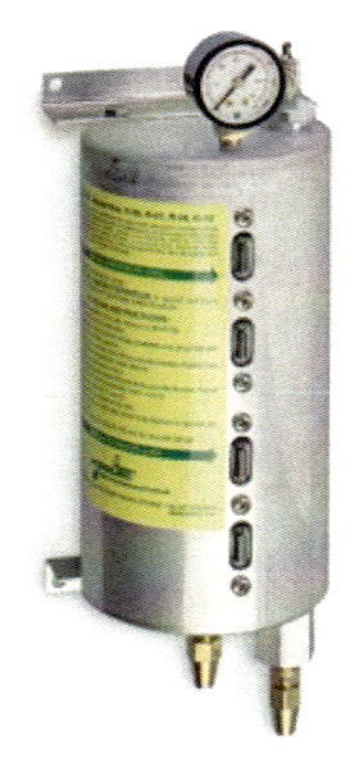

Filling and Bleeding the System

Remove fill port plug from the reservoir and fill the reservoir within one inch of the top of the sight tube with hydraulic fluid. Replace fill port plug

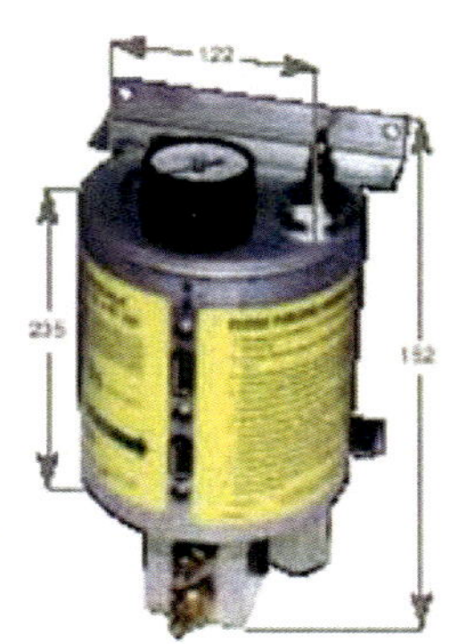

The hydraulic Fluid

During the filling procedure use the same fluid existing in the current system . If you prefer use a different fluid , remove the old fluid completely , add the new fluid and bleed the system properly

Bleeding Procedure

Each hydraulic system have your own air bleeder . Check the manufacturer specifications where is located the bleed valve and what is the procedure step by step

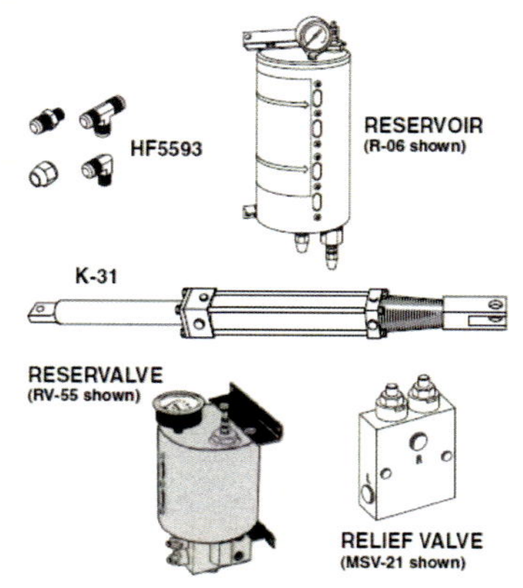

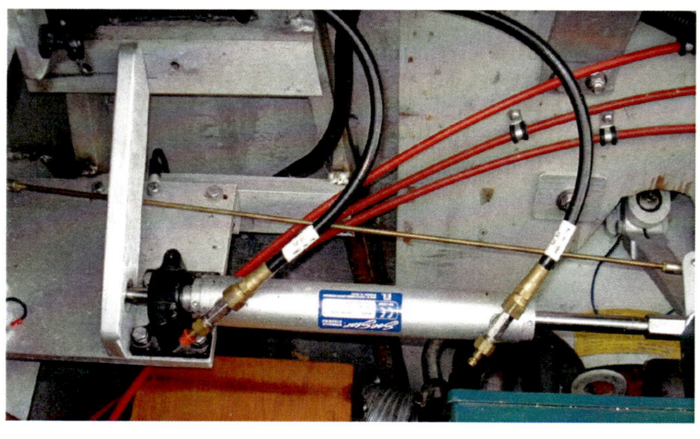

Bleeding Procedure

Part of foaming liquid will be expelled from the valve . Repeat the procedure in both directions until the liquid coming out be free of air

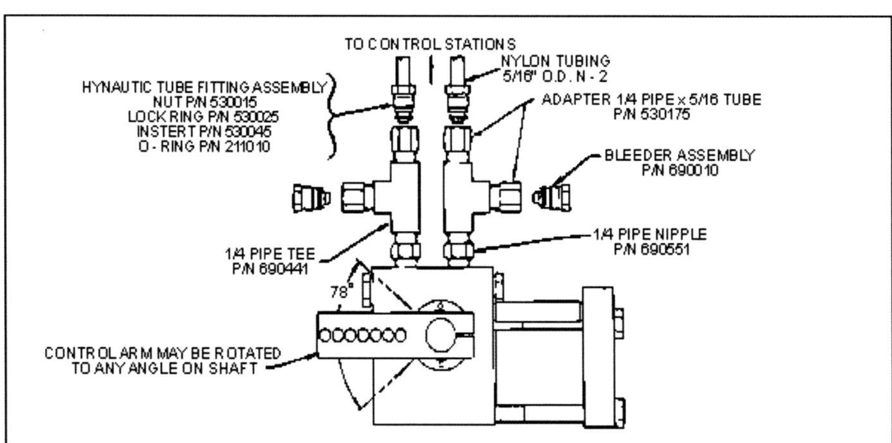

Figure 5. ST-06 Integrated Throttle Slave with Tee and Bleeder Valve Installation

Bleeding Tips

- The bleeding procedure is much easier for two people to perform than one. (One to keep the reservoir filled and pressurized, while the other one bleeds the system

- This procedure can be applied to any hydraulic system

- The reservoir must be filled at least 8 times during the bleed operation. The fluid which has been bled off should be used to refill the reservoir

Synthetic Universal Power Steering Fluid (PSF)

- Synthetic Universal Power Steering Fluid provides excellent wear protection.

- Its synthetic formulation delivers better lubricity and reduced friction, resulting in cooler operating temperatures, longer component life and quieter operation

Synthetic Universal Power Steering Fluid (PSF)

It is formulated with a high viscosity index ensures immediate lubrication at startup and in cold temperatures, yet it resists thermal breakdown and maintains maximum protection in high temperatures

Automatic transmission Fluid (<u>ATF</u>)

- The fluid is a highly specialized oil optimized for the special requirements of a high pressure and high temperature, such as body valves ,clutch packs and the torque converter as well as high pressure cylinders

- It is typically colored red or green to distinguish it from motor oil and other fluids in the Boat.

Marine <u>ATF </u>Fluid

Automatic transmission fluids typically contain antioxidants, antifoam agents, viscosity modifiers, antiwear additives, friction modifiers, and seal swell modifiers

Detergent Vs non-Detergent

- Detergent oils have the ability to emulsify water, and disperse and suspend other contaminants such as varnish and sludge.
- The main caution with these fluids is that they have excellent water emulsifying ability, which means that <u>if present, water is not separated out of the fluid</u>.
- Water accelerates aging of the oil, reduces lubricity and filterability, reduces seal life and leads to corrosion and cavitation. Emulsified water can be turned into steam at highly loaded parts of the system.

Hydraulic Oil Vs Engine Oil

- Engine oils are intended to have high resistance to heat (degradation including chemical and viscosity due to heat) resistance to burning and resistance to absorption of fuels and chemical compounds produced during combustion
- Both classes of oils are likely to have additives intended to provide detergency and to reduce foaming

Hydraulic Oil Vs Engine oil

- hydraulic and engine oils are made from base oils with additives mixed in. The additives used change the characteristics of the oils so that they function differently

  Hydraulic oils (final product including additives) are expected to have very low compressibility and very predictable friction and viscosity stability under pressure

Anti-wear vs no anti-wear

- The purpose of anti-wear additives is to maintain lubrication under boundary conditions. The most common anti-wear additive used in engine and hydraulic oil is Zinc dialkyl dithiophosphate (ZnDTP).

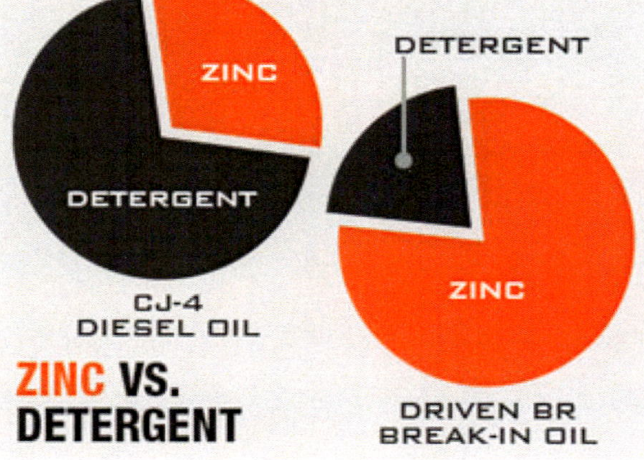

Anti-wear vs no anti-wear

It is an essential additive in any , high-performance hydraulic system, such as a boat garage doors , stabilizers and also in thruster systems

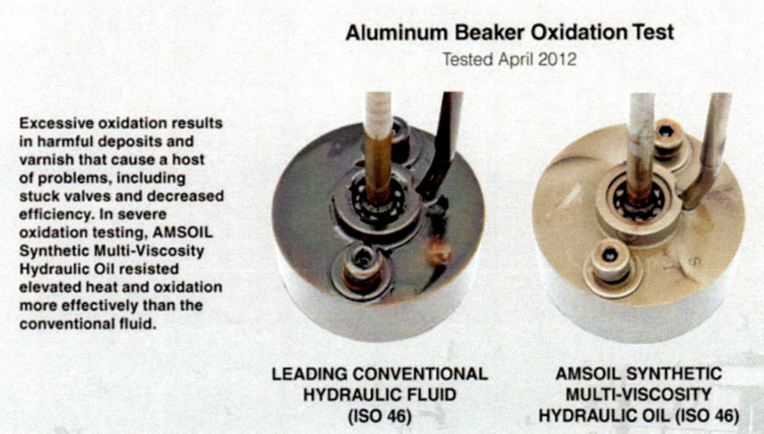

Checking the Hydraulic Fluid

- Watch out for " pink stink," the burned odor that indicates trouble inside the hydraulic unit
- If the ATF smells like burned toast and/or has a discolored brown appearance , the fluid has cooked itself and is no longer capable of providing proper lubrication to the equipment

Inspecting the hydraulic fluid

- If the fluid has a milky brown appearance, it indicates salt water contamination. There is probably a leak in the oil heat exchanger

- If the fluid is full of bubbles or is foamy, the system is probably overfilled . Other causes include using the wrong type of fluid or a plugged reservoir tank vent

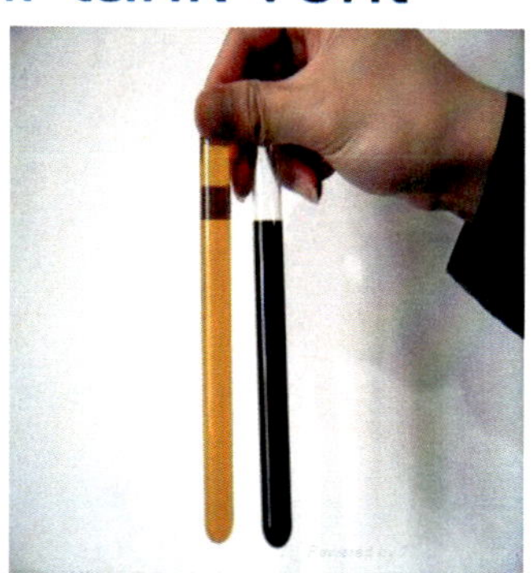

Checking the Hydraulic Fluid

- !Heat! is the main concern for hydraulic fluid. Marine systems such as thrusters, stabilizers and actuators create a lot of friction, and friction produces heat.

- For this reason all the hydraulic fluid shall pass through an oil heat exchanger to keep operating temperatures

Cooling the Hydraulic Fluid

- Most hydraulic fluids can withstand normal operating temperatures of around 200 degrees F for thousands of hours. But if the temperature of the fluid rises above 220 degrees F the fluid starts to break down quickly

- Above 300 degrees, fluid life is measured in hundreds, not thousands of hours. And above 400 degrees, the fluid can self – destruct in <u>20 to 30 minutes</u>

Checking the Fluid Level

- If the fluid level is low , the hydraulic equipment may slip or engage slowly. If the level is too high, the fluid can become mixed with air (aerated) causing shifting problems, slippage and noise

- A low level , therefore, usually indicates a leak. A visual inspection of the pan gasket and driveshaft seals will tell you where the fluid is

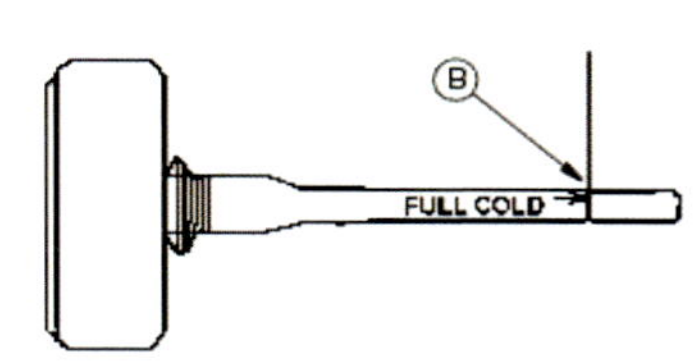

Flow Meters

As an electrical circuit, the Flow-meter shall be inserted <u>in series</u> with the fluid lines as the Amp-meter in the electrical circuit

The measurement units are: Gallons per hour (GPH) or Gallons per minute (GPM)

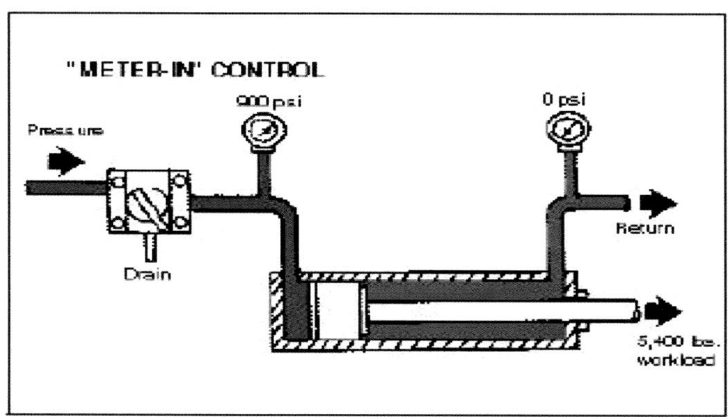

Hydraulic Schematic Diagram

A schematic diagram is a 'road map' of the hydraulic system and to a technician skilled in reading and interpreting hydraulic symbols, is a valuable aid in identifying possible causes of a problem. This can save a lot of time and money when troubleshooting hydraulic problems

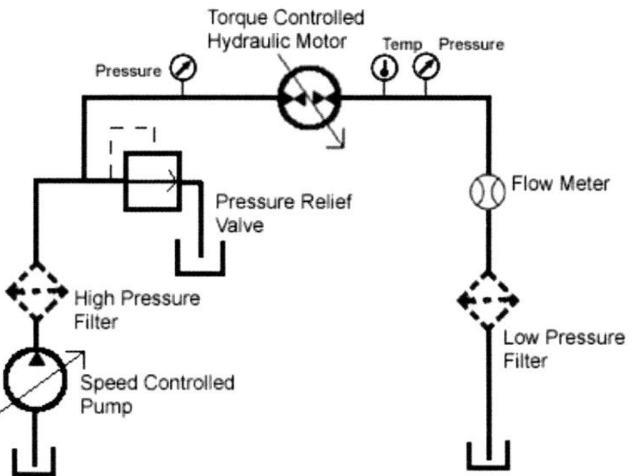

Hydraulic Schematic Diagram

- If a schematic diagram is not available, the technician must trace the hydraulic circuit and identify its components in order to isolate possible causes of the problem. This can be a time-consuming process, depending on the complexity of the system.

Hydraulic-Circuit Diagrams

- The four types of hydraulic-circuit diagrams are block, cutaway, pictorial, and graphical.

- **Block Diagram.** A block diagram shows the components with lines between the clocks, which indicate connections and/or interactions.

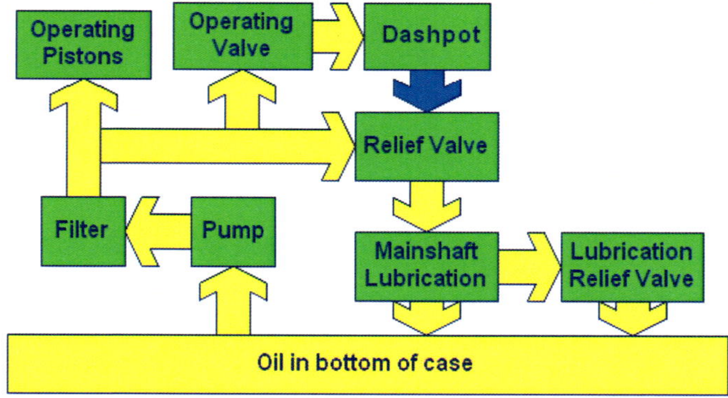

Hydraulic-Circuit Diagrams

- **Cutaway Diagram.** A cutaway diagram shows the internal construction of the components as well as the flow paths. Because the diagram uses colors, shades, or various patterns in the lines and passages, it can show the many different flow and pressure conditions.

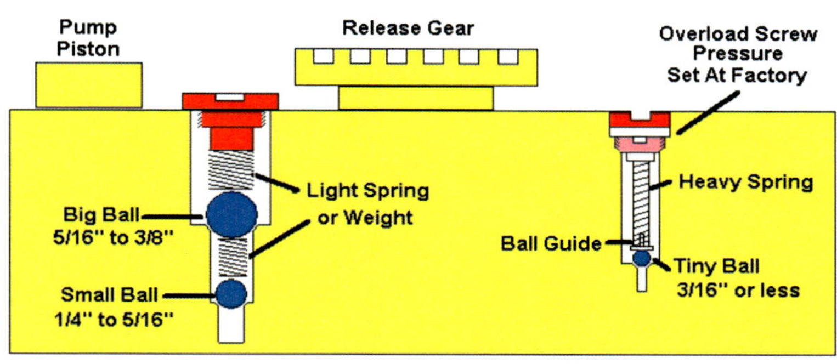

CUTAWAY VIEW OF JACK BLOCK

Hydraulic Circuit

- Schematic

- Hydraulic Diagram

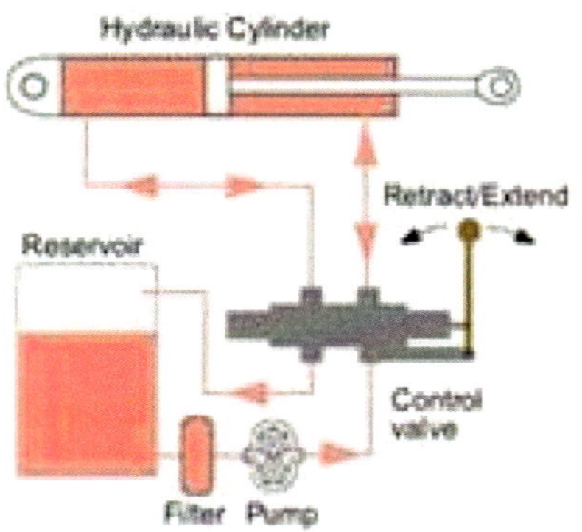

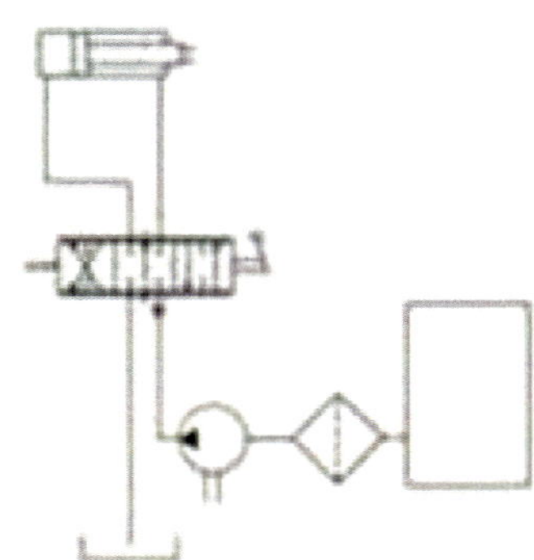

Hydraulic Basic Symbols

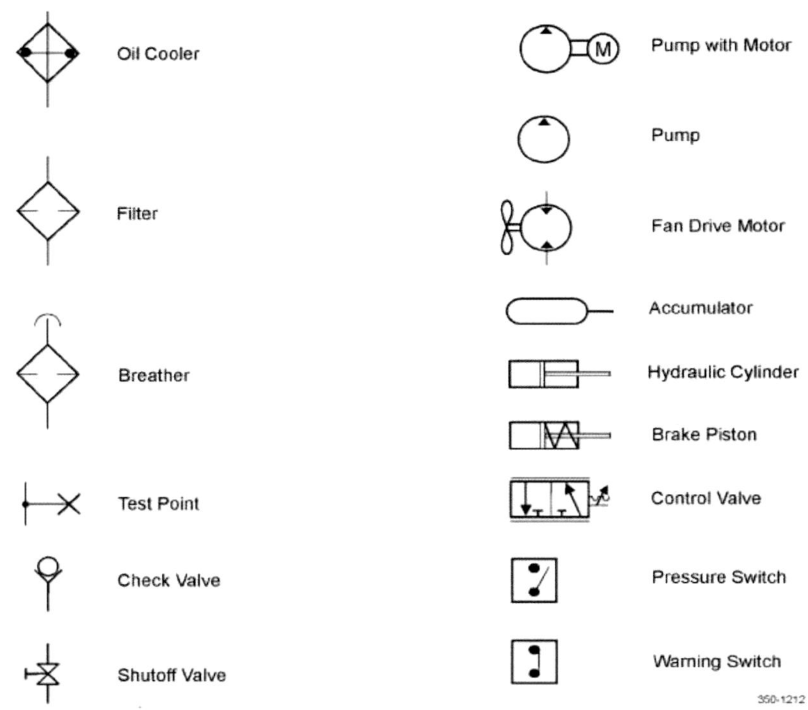

Hydraulic Valve Operator Symbols

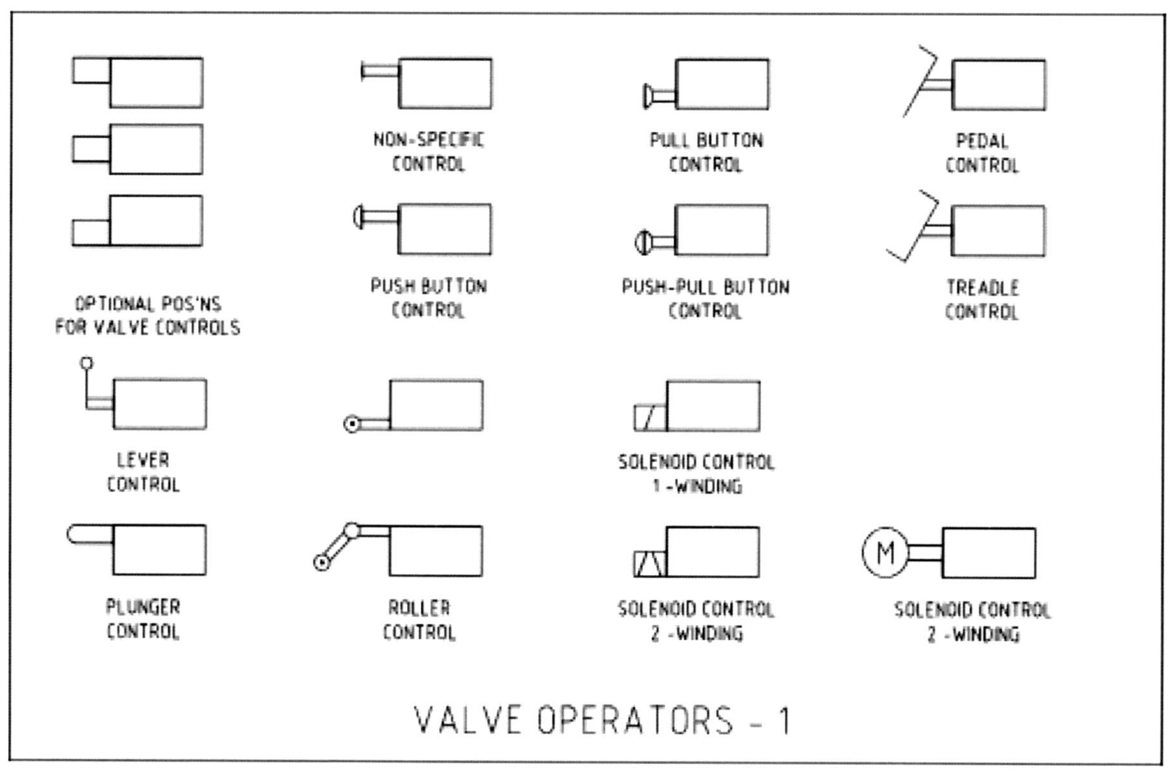

Readings Instruments

- The fluid meter senses the amount of liquid or gas that is passing through the pipe during an specific fraction of time

- The measurement units are: GPM and GPH

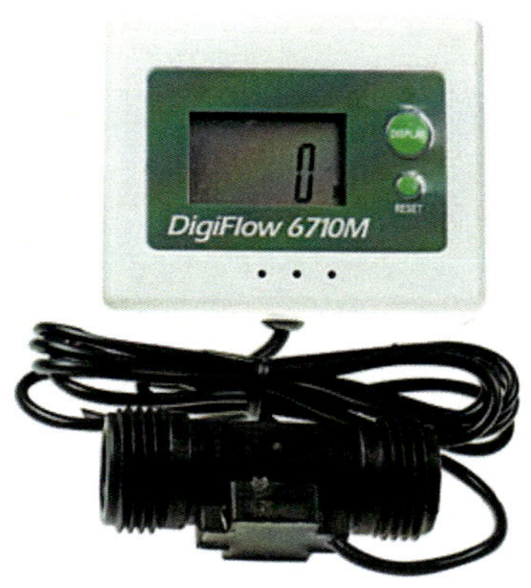

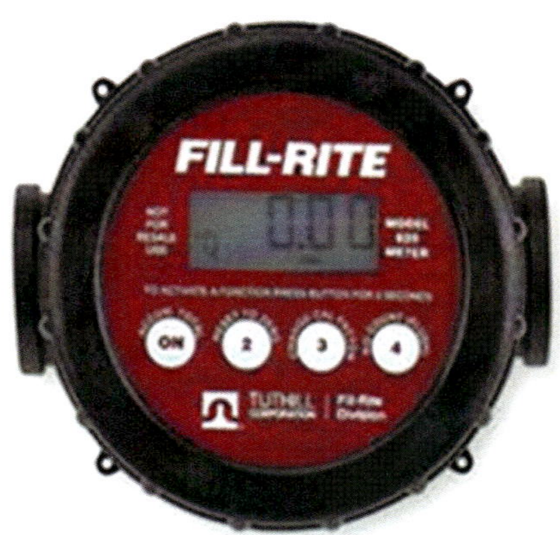

Flow Meters

A flow meter is an instrument used to measure the flow rate of a liquid or a gas

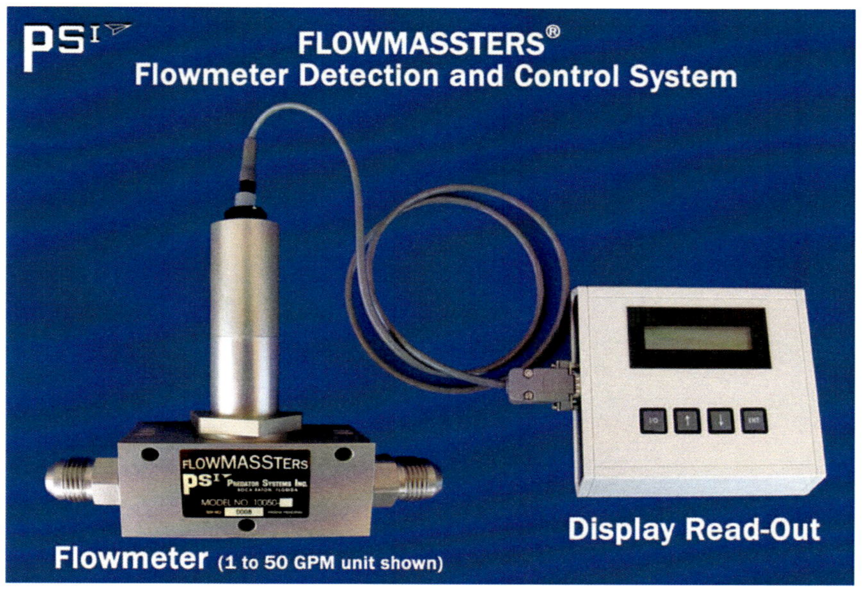

Flowmeter (1 to 50 GPM unit shown)

Chapter 3
Trim Tabs

- Trim tabs Specification
- Trim Tabs Operation
- Hydraulic Unit
- Wiring a Trim Tabs system
- Troubleshooting
- Trim Tabs System with Interceptor
- Selecting a Trim System

Trim Tabs

Trim tabs are small control surfaces or stabilizers connected to the transom of the boat

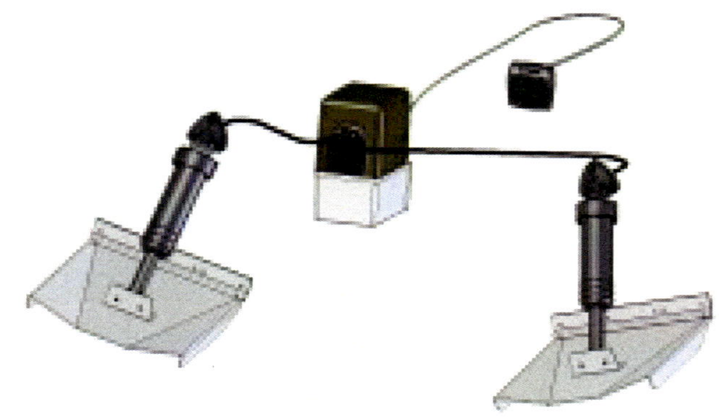

Boat Trim Tabs

Large and small boats benefit from trim tabs for the same reasons—they get the boat out of the hole and on plane quickly, they correct for uneven weight distribution, improve speed, safety, and overall boat performance

Boat Trim Tabs

Trim tabs usually consist of two adjustable stainless steel planes mounted at the transom of the boat. Controlled by a hydraulic power unit, the tabs can move up and down when activated by the boater

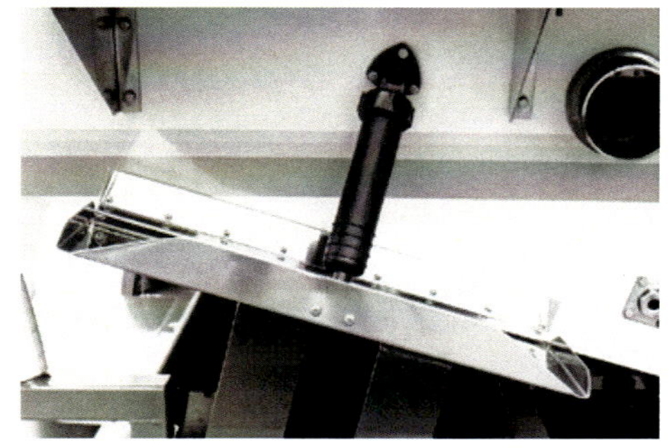

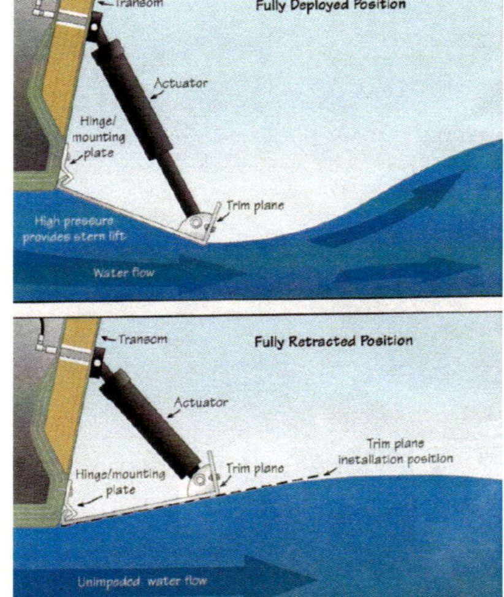

The Benefits of Adding Trim Tabs

- **Performance**—Trim tabs increase speed, reduce pounding, correct listing, eliminate offset prop torque
- **Efficiency**—Reduce fuel consumption, reduce engine laboring, and eliminate squatting
- **Safety**—Improve visibility, reduce wake size, improve handling, and reduce hull stress

Getting on Plane

- In order to get on plane faster, boaters often have to ask passengers to move forward. With the additional lift from the trim tabs, the boat will spend less time operating in the inefficient transition period before planning
- The engine labors less, the boat gets better fuel economy, and passengers can sit where they please

Independent Control

- Because trim tabs are mounted on both sides of the vessel's stern and can be operated independently, the vessel's side to side trim may be controlled by deflecting one side more than the other

- This independent control of the tabs is essential in correcting for port or starboard lists

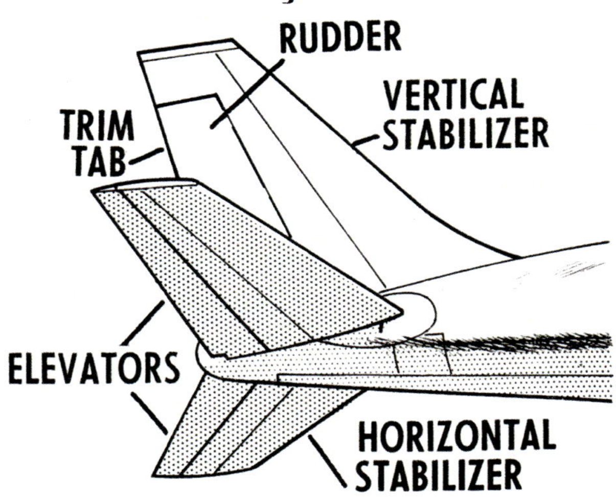

Aircraft Trim Tabs

Tabs used to trim any out-of-balance forces and permit a hand-off flight at normal speeds. These tabs are mounted on aircraft primary controls and can be operated from the cockpit either by hand wheels or electrically. Also called trimmers

Trim Tabs

- Boats with planning-type hulls will often have trim tabs attached to the trailing edge of the hull or transom.

- Changes in boat speed or weight placement will usually require the trim tabs to be adjusted to keep the boat at a comfortable and efficient pitch attitude. This reduces the work of the captain by reducing the amount of manual control necessary

Trim Boat and Engine Strain

- In addition to increased fuel efficiency, Trim Tabs adjust to your boats ideal cruising attitude resulting in less strain on your engine. This improved attitude also reduces the pounding that results from riding in rough water, or uncomfortable proposing.

- Trim Tabs also gives you increased performance from your prop. An untrimmed prop creates inefficient, turbulent path resulting in both slippage and decreased speed and performance

Trim Boat and Engine Strain

- With Trim Tabs, your prop propels the boat at peak efficiency creating maximum thrust and a straight, forward motion.

- By adjusting your boat to the proper attitude, your R.P.M.'s will increase in the throttle. You can trim your craft to the most efficient planning angle for a ride you never thought was possible for your boat.

- To correct listing problems resulting from the wind, prop torque or the weight distribution of your passengers simply adjust each Trim Tab independently.

Hydraulic Unit

Is a compact powerhouse unit that creates hydraulic pressure to lower and raise the trim tabs. It employs two solenoid valves to direct the flow of fluid to the actuators

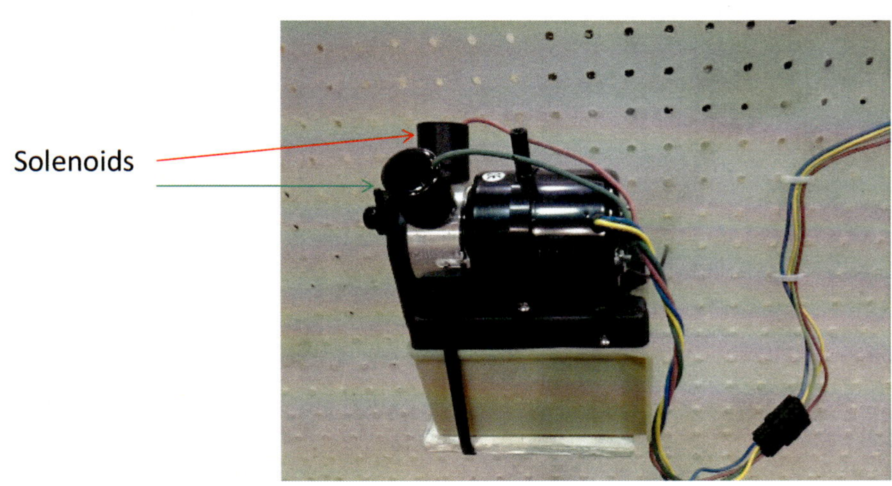

Hydraulic Pump Schematic

The motor turns clockwise once the yellow cable receive the power and it turns counterclock once the blue cable is powered

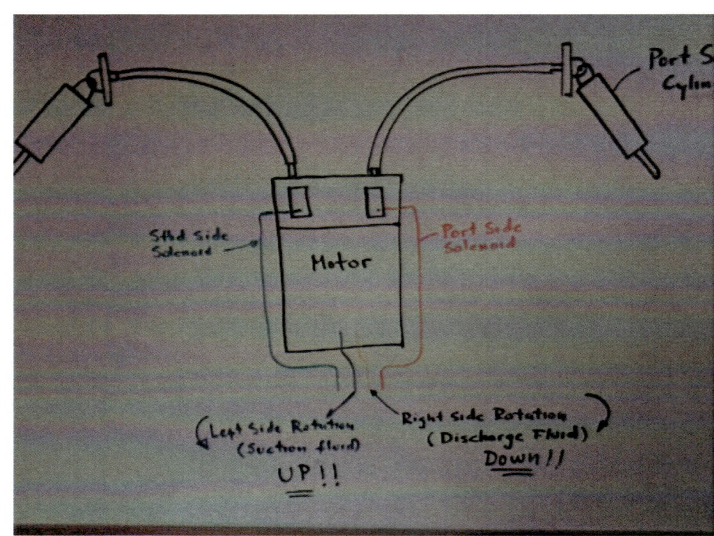

Trim Tabs Kit

Standard Trim Tab Kits include: (2) actuators, (2) 12 gauge S.S. trim tabs, and all mounting hardware (Switch sold separately).

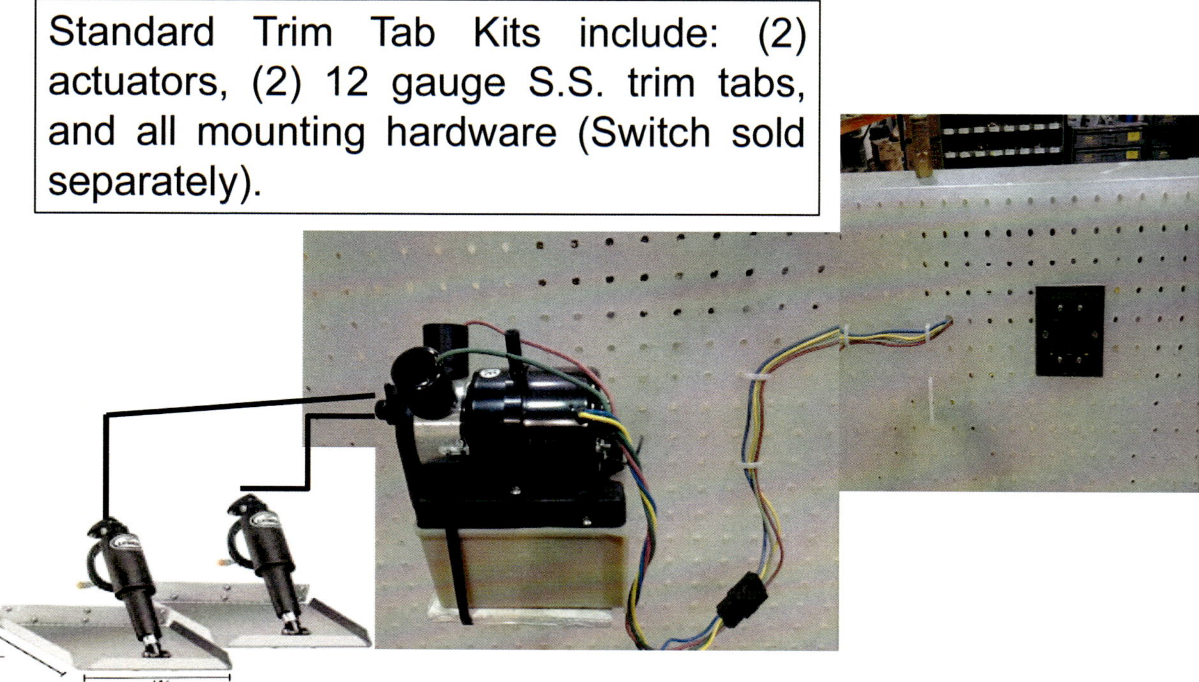

Trim Tabs Switch

- The waterproof toggle switches are the best in the industry
- The switches come with all mounting hardware and a mounting gasket
- The resettable relay allows the operator to manually reset the relay in the event it has tripped due to operator error or malfunction of the unit

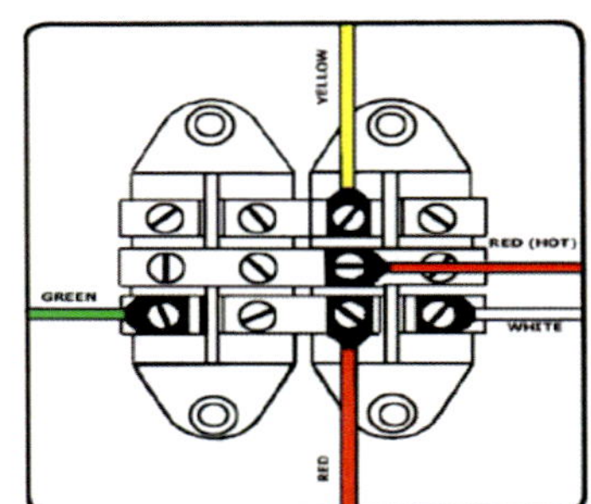

Hydraulic   vs.   Electric

- Hydraulic systems are capable of making the precise adjustments needed for smooth safe control of the trim tabs
- Hydraulic systems start and stop instantaneously. There is no lag time

- Electric motor create considerable lag time from the time you take your finger off the control until the trim tab stops moving
- This "actuator overrun" means that even momentary presses of the switch will result in the trim tabs running too far down or too far up

Hydraulic   vs.   Electric

- A failure in a hydraulic system does not result in locked up trim tabs. Then simply removing pressure from the system will result in the trim tabs retracting to the full up position. With the trim tabs in the full up position, the boat may safely run at high speeds

- When an electromechanical system fails, the actuator will lock the trim tab in place. If the trim tab is in the deflected position when a failure occurs, the only way to retract it is to haul the boat out of the water and replace the actuator

Electronic Indicator Control

- Three features combined into one sophisticated control ; switch control, trim tab indicator and the Auto Tab Retractor

- Water resistant, highly accurate and features variable intensity LEDs for night or day

Hydraulic Cylinders (Nylon)

The hydraulic cylinders are injection molded from glass-filled nylon material. With the upper bracket pivoting approximately 270 degrees this cylinder can be mounted at most any angle

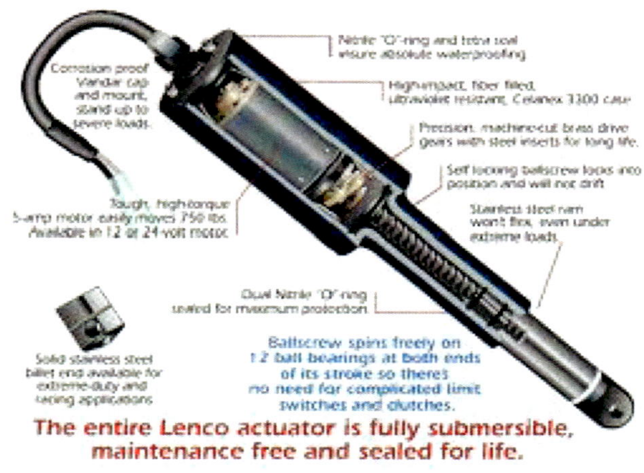

Hydraulic Cylinders (Bronze)

Some boat manufacturers include a full flap across the transom .Because the big area of the flaps, a group of 8 actuators are required to manipulate the flaps

Hydraulic Cylinders (Bronze)

Due to the high pressure required to move the flaps the cylinders are constructed in bronze alloys and non corrosive steel alloys

Trim System with Interceptors

The Boat Trim System (BTS) with its patented interceptor technology gives you perfect control over pitch and heel with rapid response, quicker onto plane, lower fuel consumption and a more comfortable ride

QL Boat Trim System

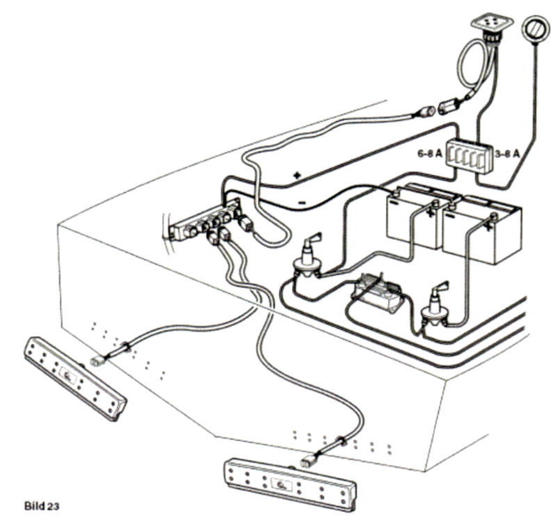

Bild 23

Trim System with Interceptors

Unlike conventional trim systems using hydraulics the system is operated electronically. As no hydraulic oil is used, there's no risk of oil leakage in sensitive marine environments

Trim System with Interceptors

The trim system, recommended for speeds up to 45 knots, has all the same performance qualities as conventional trim tabs plus the additional benefits of quicker response, less drag and compact design

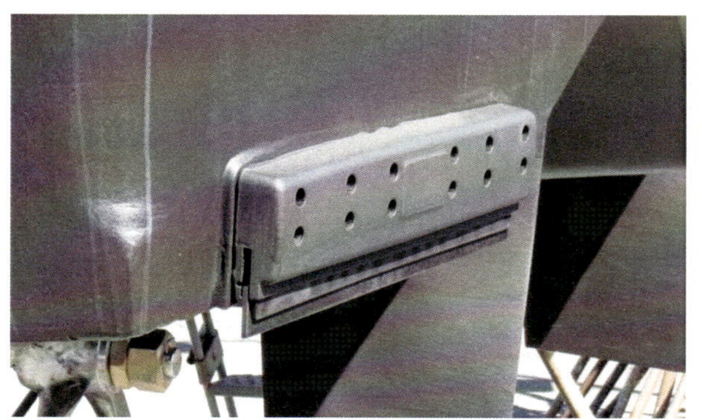

Power Trim & Steering Fluid

- The hydraulic fluid that transmits the power in your Boat Trim Tabs and power steering system is called as power-steering fluid.

- The fluid is the cheapest component of your power-steering system However it is this fluid that steers your Boat safely out of any trouble.

- If your power steering fluid does so much the least you could do is to replace it by fresh power steering fluid on schedule.

Chapter 4
Windlass & Anchor System

- Electric Windlass Components
- Selecting the Windlass
- Wiring & Schematic Connections
- Windlass Configuration
- Sizing the Windlass
- Rope & Chain Selection

Electric Windlass Components

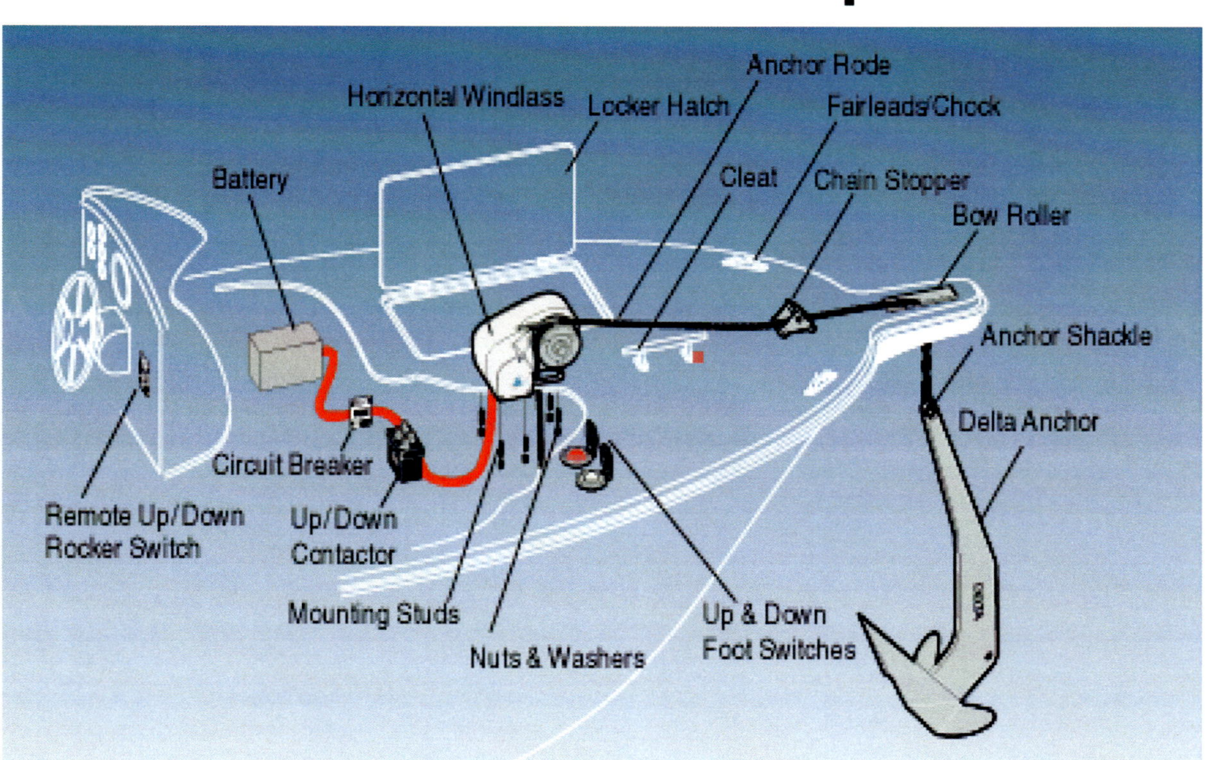

Electric Windlass components

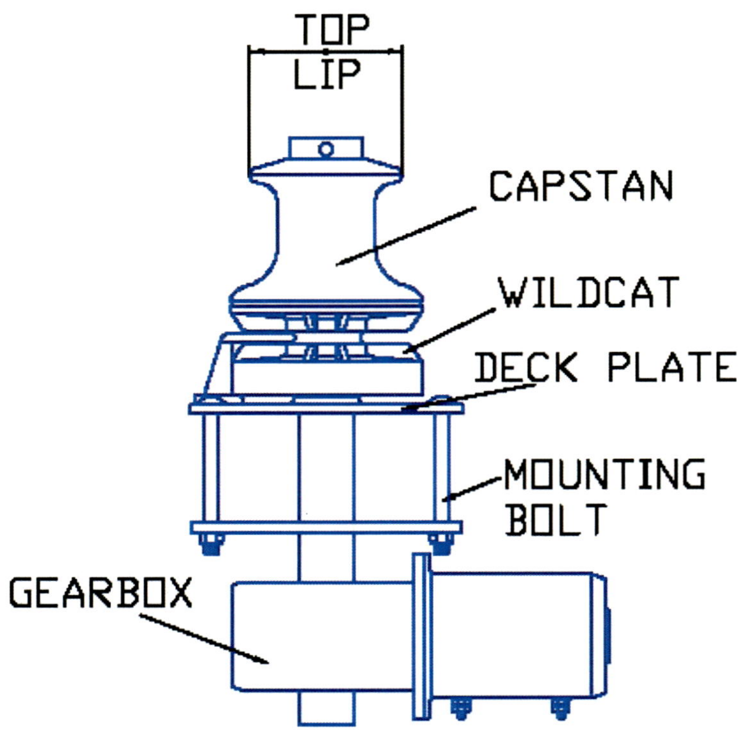

Selecting the Windlass

- There are a number of important criteria to be considered in selecting the correct anchor *winch*

- The vessel size, displacement, windage, anchor size and *rode* selection

- Practicalities such as locker space and depth of fall for the rode also play a part in deciding which *windlass* is ideal for you

The Electrical Motor

- The electrical motors are reversible motors of 12 VDV or 24 VDC

- Normally the 12 Volt motors consume 40 amps without load and around 140 amps with load

The Electrical Motor

- The lifting capacity of this type of motors is between 300 Kg and 400 Kg

- Holding capacity at anchor is 3500 kilos

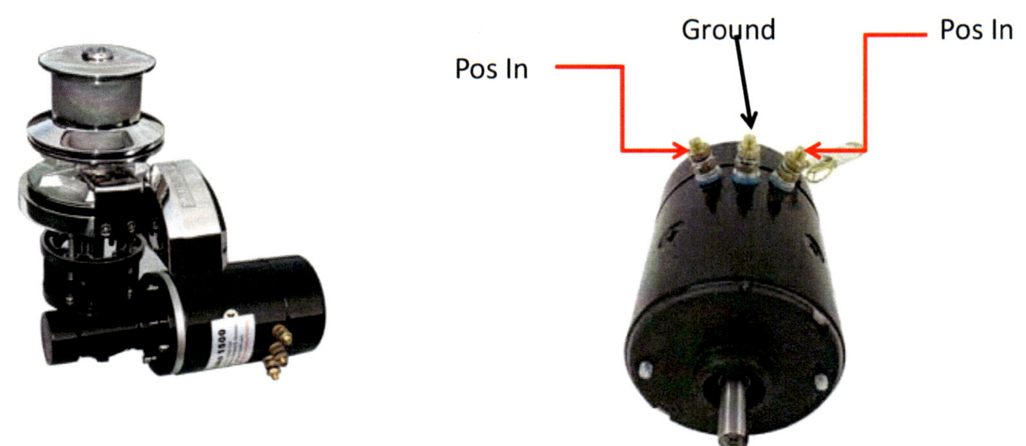

The Windlass Battery bank

- The electric windlass should run with your own deep cycle battery bank , with an ampacity according with motor manufacturer recommendations plus 20% (safety factor)
- In order to design an efficient system try to install the batteries close to the electric motor to avoid excessive voltage drop

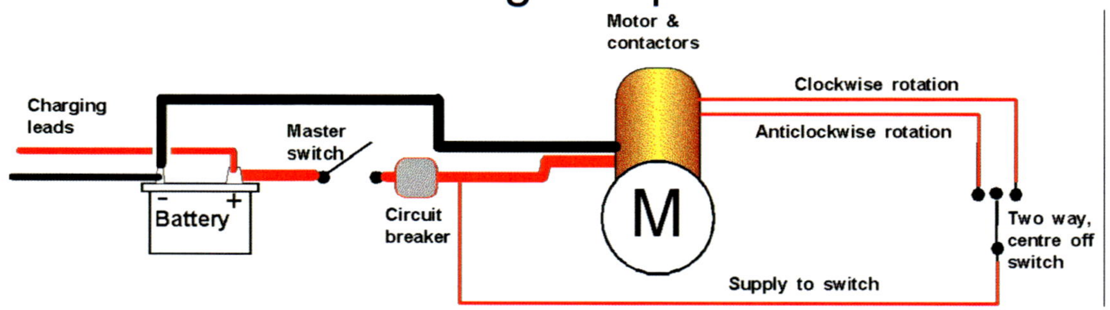

The Electrical Motor

Normally the motor have three terminals. The middle one is connected to the ground and the lateral posts receive the power from the solenoid control box

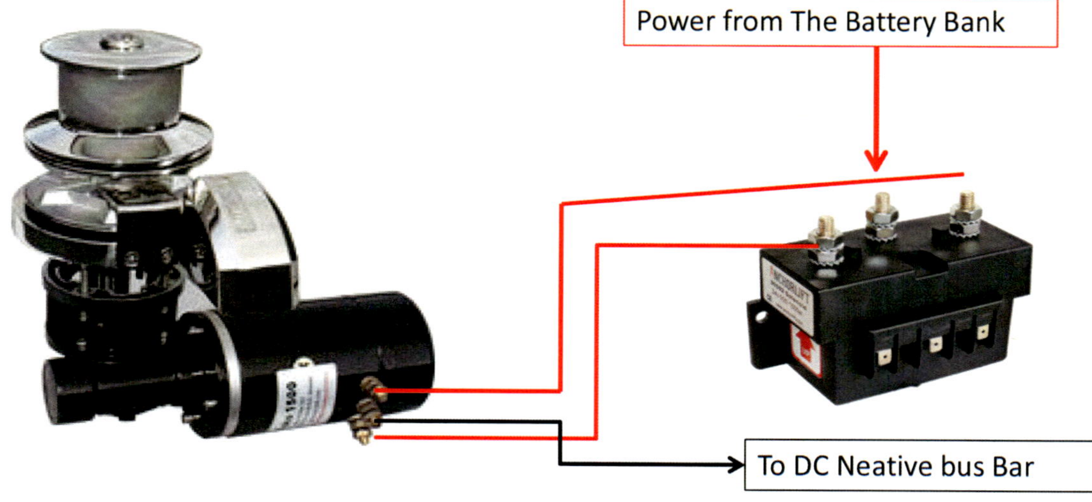

Sizing The Wires

The size and length of the cables depend of the distance between the batteries and the electric Motor. To calculate properly those dimensions I recommend use the tables 1 & 2 in page 53 at the ***Marine Electricity handbook***

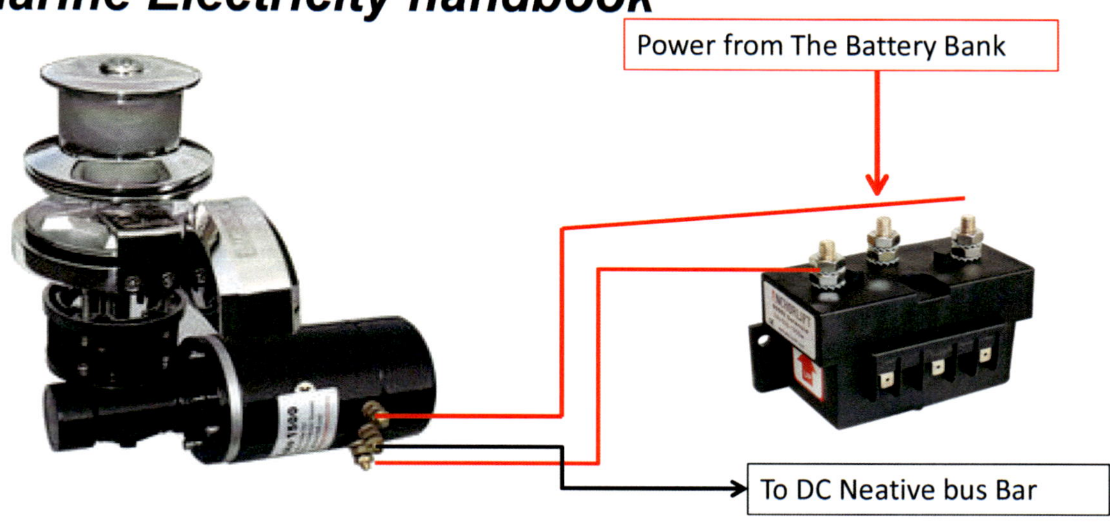

The Solenoid Control Box

It is the connection between the Electric motor with the foot-swithes and rocker-switch

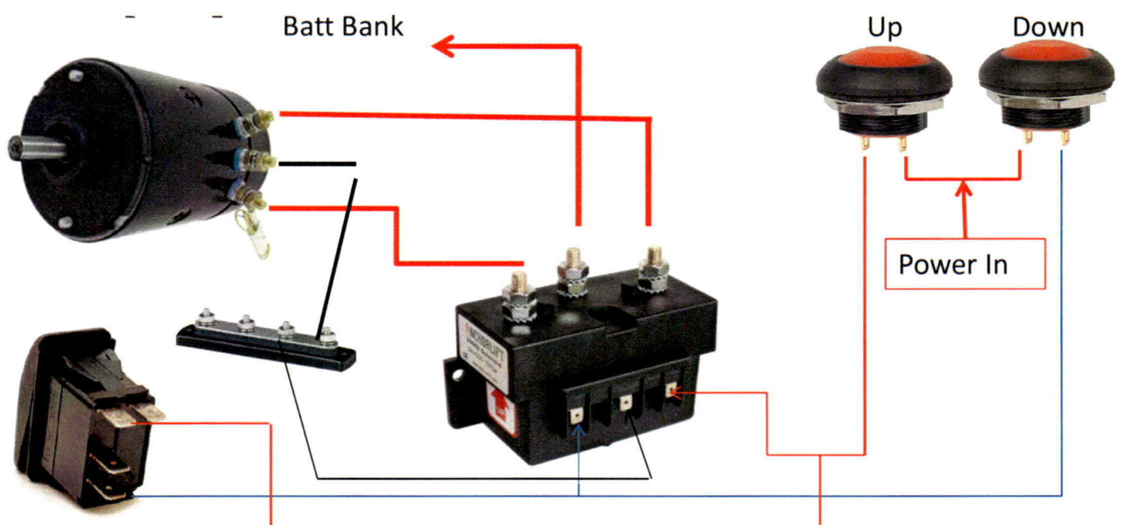

The Solenoid Control Box

Due to the high cost of the solenoid control box , it can be replace for a couple of cheap automotive relays

Windlass Wiring Diagram

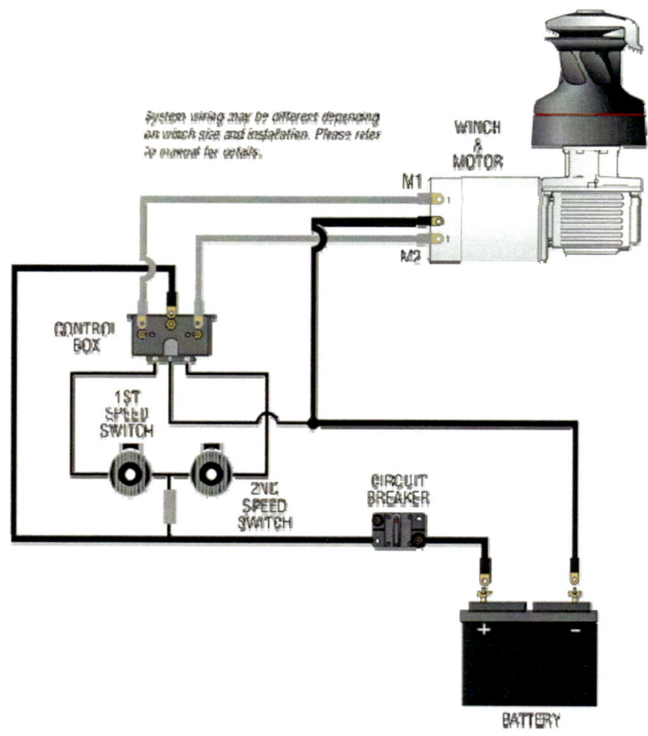

Vertical or Horizontal Configuration

- The two basic types of windlasses are differentiated by the drive shaft orientation

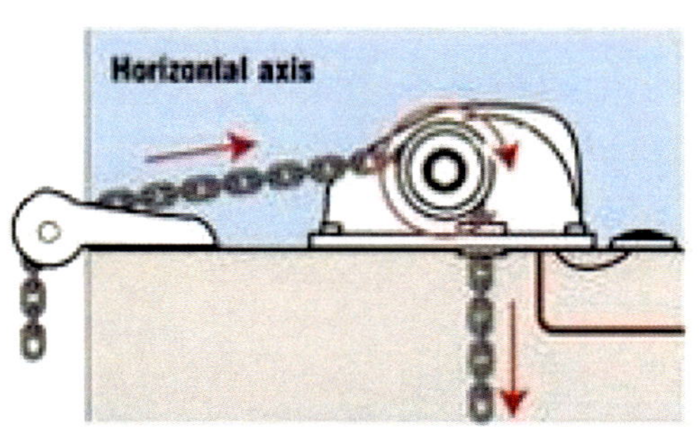

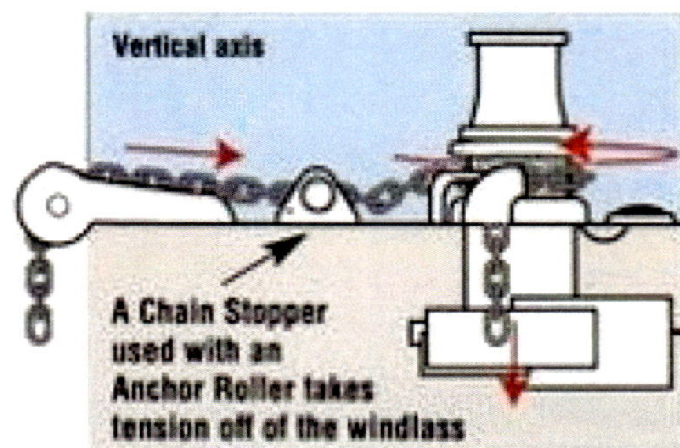

- Deck thickness and underdeck space are the two main considerations when deciding which of the two types to fit

Vertical Configuration

Vertical windlasses are characterized by situating the *capstan* and/or **gypsy** above deck and the motor and gearbox below deck

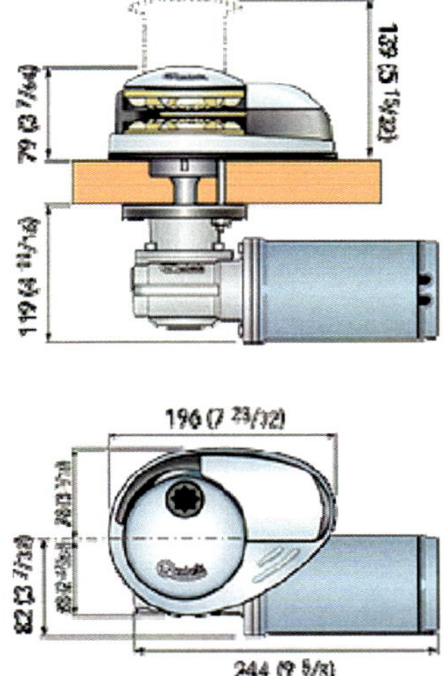

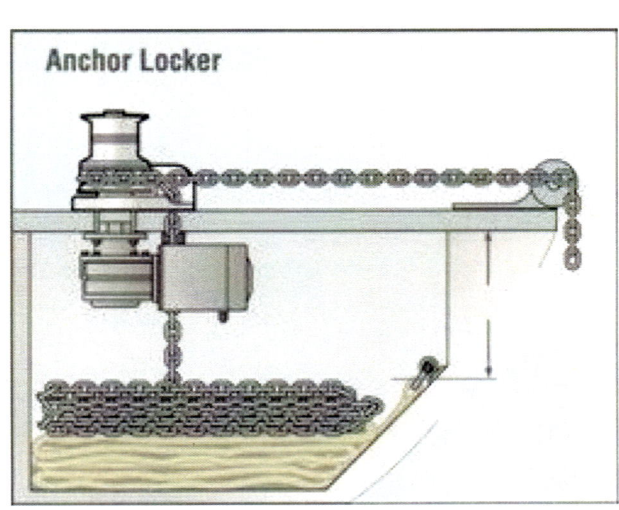

Vertical Configuration

Vertical windlasses provide a 1800 wrap of the anchor rode around the chain-wheel giving optimal chain control, minimizing slippage and jumping

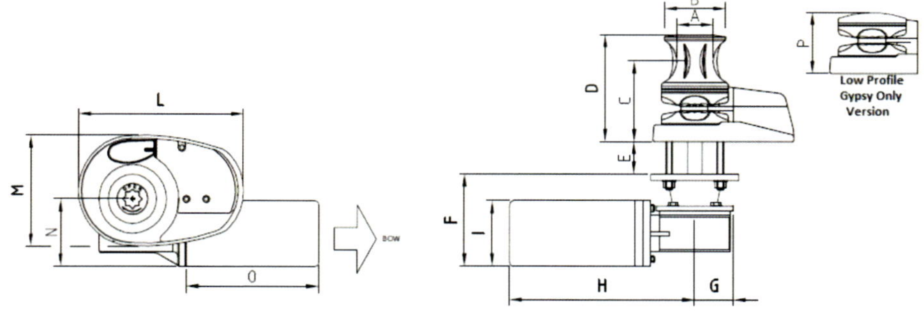

TYPE	A	B	C	D	E	F	G	H	I	L	M	N	O	P	
THUNDER 700	76	110	139	180	25/50	149	68	245	95	250	165	117	175	118	
THUNDER 1000	76	110	139	180	25/50	149	68	245	114	250	165	117	175	118	
THUNDER 1400	76	110	139	180	25/50	149	79	280	114	250	165	117	195	118	

Horizontal Configuration

Horizontal windlasses are mounted completely above deck with **gypsy** and **capstan** located to either side. They provide a 900 wrap of the rode around the chain-wheel

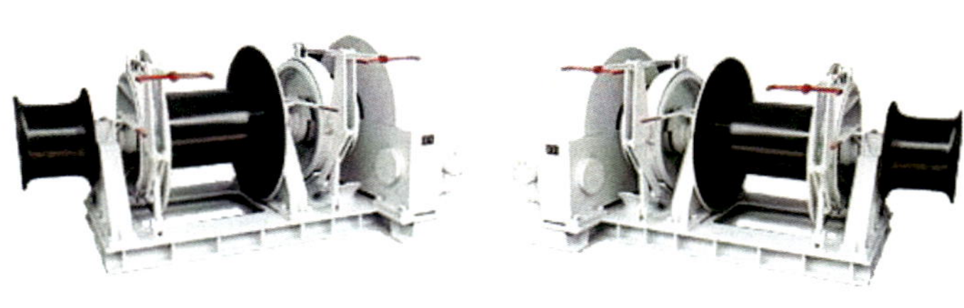

Horizontal Configuration

Horizontal models has the advantage of being better suited to applications where the accessibility is limited below deck space , extreme deck thickness (over 200mm - 8"), or where two anchors must be handled from one winch

HOW MUCH SPACE DO I NEED IN MY CHAIN LOCKER?

Deck thickness and locker space play an important role in deciding whether to install a *vertical* or *horizontal* windlass

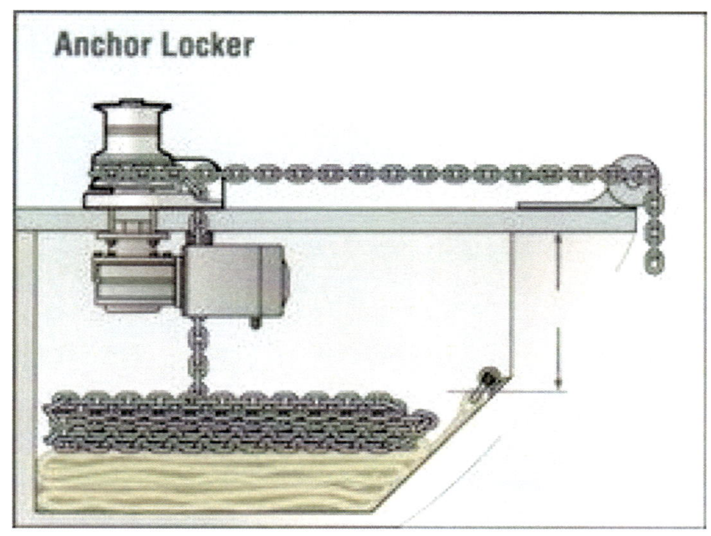

HOW MUCH SPACE DO I NEED IN MY CHAIN LOCKER?

- Estimating or measuring the depth of fall of the rode into the anchor locker may dictate which type of windlass is most suitable for your vessel

- Calculating the depth of fall differs for horizontal chain only windlasses and for vertical rope or rope/chain windlasses

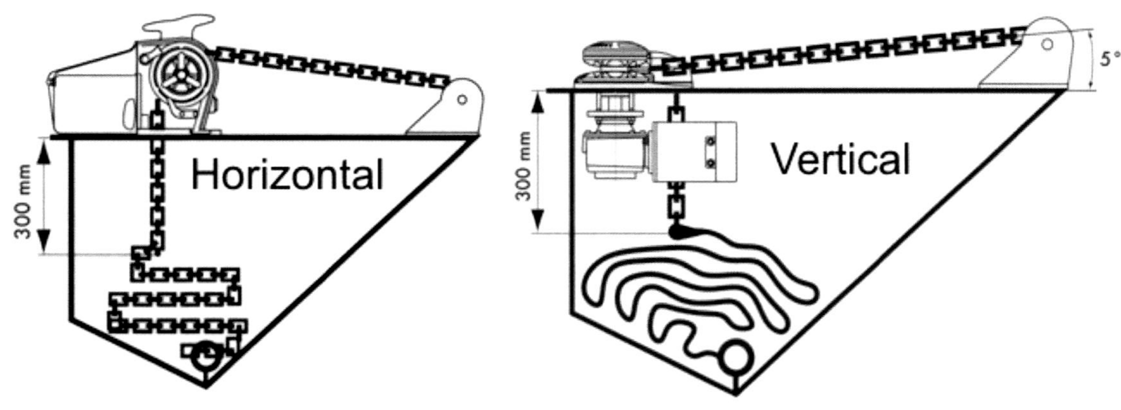

DC / AC or Hydraulic?

- With the increasing popularity of powerful and compact on-board generators, AC powered winches are becoming a practical consideration for bigger boats

- Hydraulic systems provide another power source well worth considering as they have the advantage of constant speed under all load conditions and can be run almost constantly while coupled with safe guards such as pressure relief valves

- Modern hydraulic systems offer an integrated, low maintenance and efficient, centrally managed, power pack

How to Size a Windlass

- Calculate the weight of your ground tackle. Use a scale to weigh your anchor, chain and rope

- Multiply the weight of your ground tackle by three.
    - **F = W** X **3** = Pulling Capacity.
- The result from the two previous steps is the minimum safe pulling capacity for your windlass

Rope / Chain Selection

- Rope and/or chain, particularly chain selection, is extremely important. Deciding on the right anchor winch for your boat depends on the size, not only of the boat, but also the ground tackle
- Automatic rope/chain systems are now commonly used on boats up to 20 meters (65 feet)

Types of Chain

- Short link chain is most commonly used on small and medium sized boats while stud link chain is generally used on much larger vessels such as Superyachts

- The latter is characterized by a stud (bar) joining the two sides of the link preventing them from deforming when overloaded

Chain Selection

- It is important that the right size and right grade of chain is used to ensure a correct fit of the links to the *gypsy*
- If the chain is not matched to the chain-wheel problems may occur, such as the chain jumping off the gypsy or the chain jamming as it will not feed smoothly through the chain pipe

Sizing an Anchor Chain

- Determine the length of your boat.

- Determine the weight of your boat (Full loaded)

- Use nautical charts to determine the estimated depth of waters at your anchorage site or sites

Sizing an Anchor Chain

- Determine the diameter of chain to use. Use the following chart to determine the minimum diameter of chain to use, for your boat.
- Boat Length: Boat Weight: Chain Diameter:
  20 to 25 feet ----- 2,500 lbs. -----   3/16-inch
  26 to 30 feet ----- 5,000 lbs. -----   1/4 -inch
  31 to 35 feet ----- 10,000 lbs. ----- 5/16-inch
  36 to 40 feet ----- 15,000 lbs. ----- 3/8-inch
  41 to 45 feet ----- 20,000 lbs. ----- 7/16-inch
  46 to 50 feet ----- 30,000 lbs. ----- 1/2-inch
  51 to 60 feet ----- 50,000 lbs. ----- 9/16-inch

Sizing an Anchor Chain

- Determine the length of chain to use. Your total "anchor rode" (the nylon rope and chain combined connecting the anchor to the boat) should be between four and seven times the expected anchorage depth, depending on how crowded the anchorage

- More crowded anchorage sites typically mean using a shorter rode, because boats don't have as much room to swing

Sizing an Anchor Chain

- To determine the chain portion of the rode, again refer to the length of your hull

- Your chain should be no shorter than the length of your boat, as the chain's weight will help set the anchor

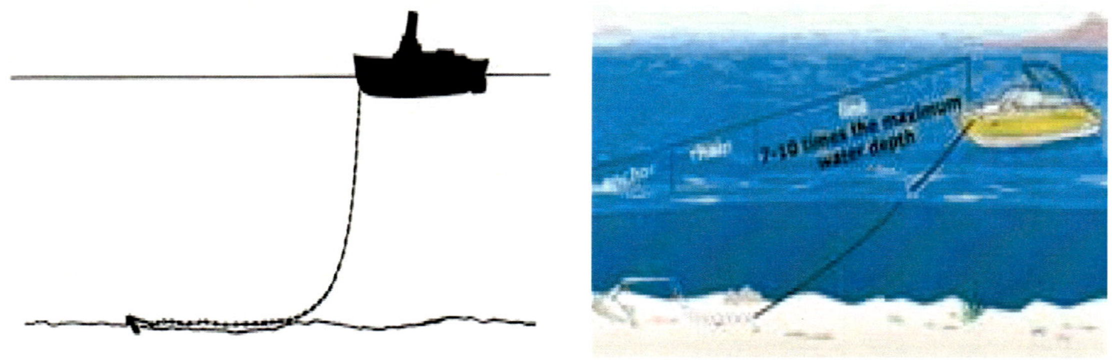

Chapter 5
Power Steering Systems

- Rack and Pinion Steering
- Rotary Steering Helm
- Troubleshooting Mechanical System
- Hydraulic Steering System
- Power Assisted System
- Auto Pilot System
- Preventive Maintenance System

Marine Power Steering

- The steering of your boat is likely handled either **by hydraulics or by a mechanical** steering system

- Mechanical systems are typically either rack and pinion or rotary, which is sometimes called cable.

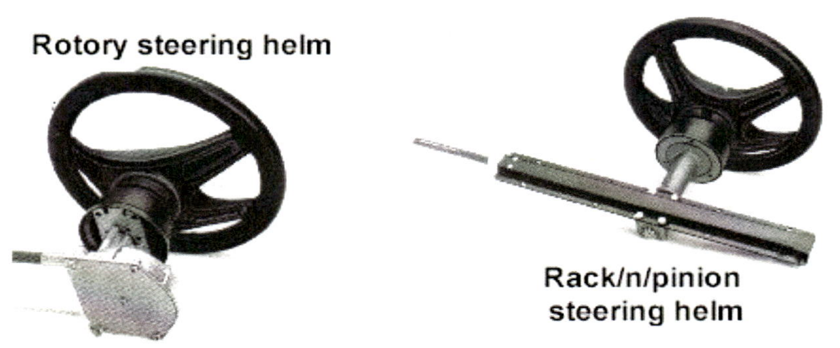

Rack and Pinion Steering

Rack and pinion steering employs a rack that is moved back and forth by a pinion. The steering cable is attached to the rack, which pushes and pulls the steering cable

Rack-and-pinion Steering

A rack-and-pinion gearset is enclosed in a metal tube, with each end of the rack protruding from the tube. A rod, called a **tie rod**, connects to each end of the rack

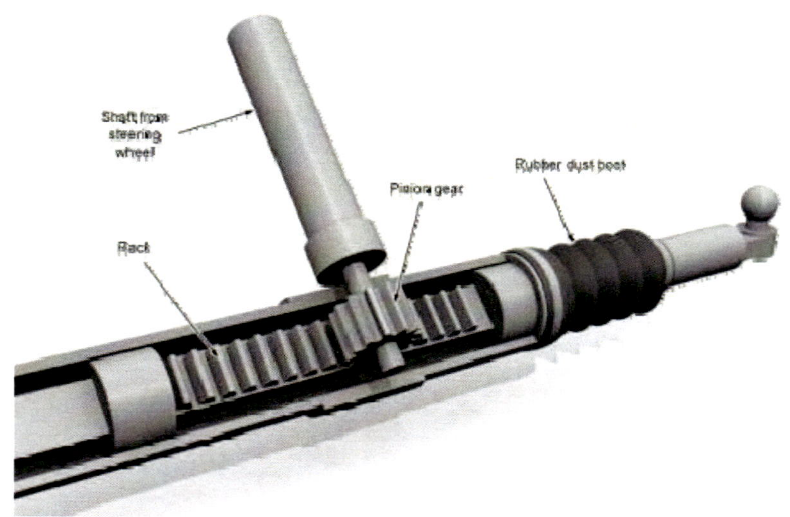

Rack-and-pinion Steering

The **pinion gear** is attached to the **steering shaft**. When you turn the steering wheel, the gear spins, moving the rack. The tie rod at each end of the rack connects to the **steering arm**

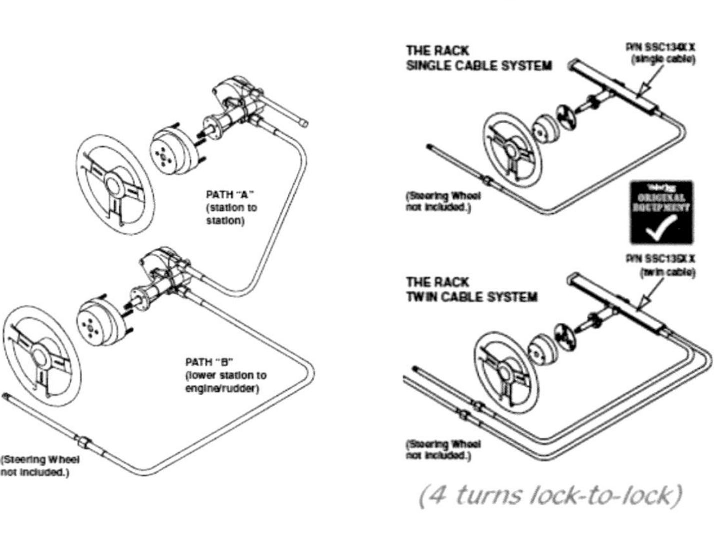

(4 turns lock-to-lock)

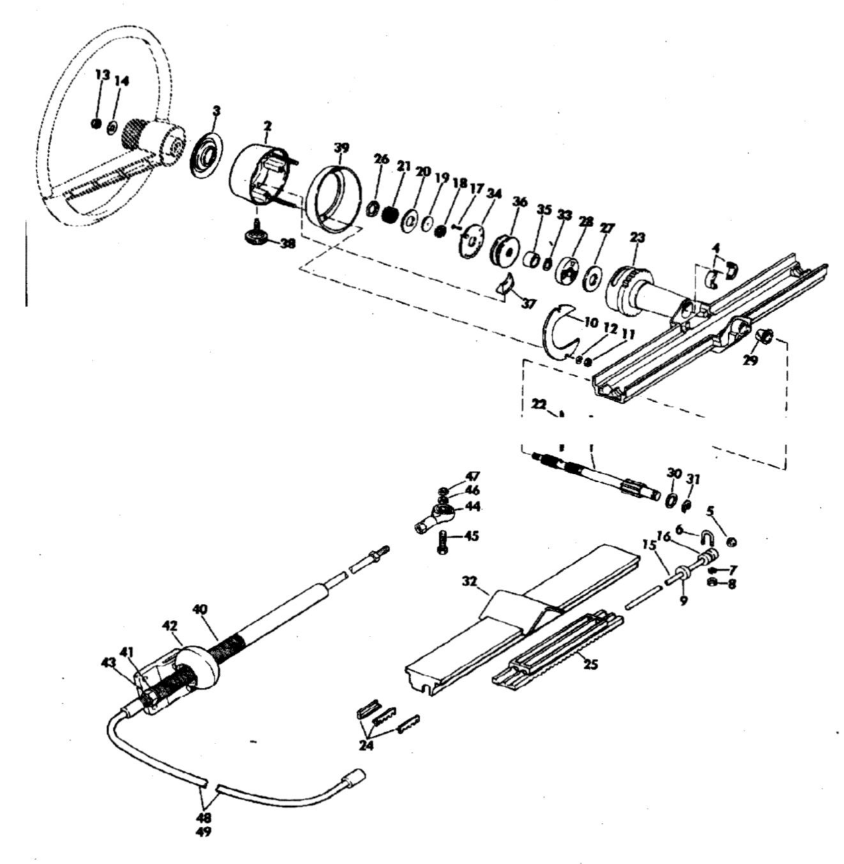

Rack and Pinion Steering

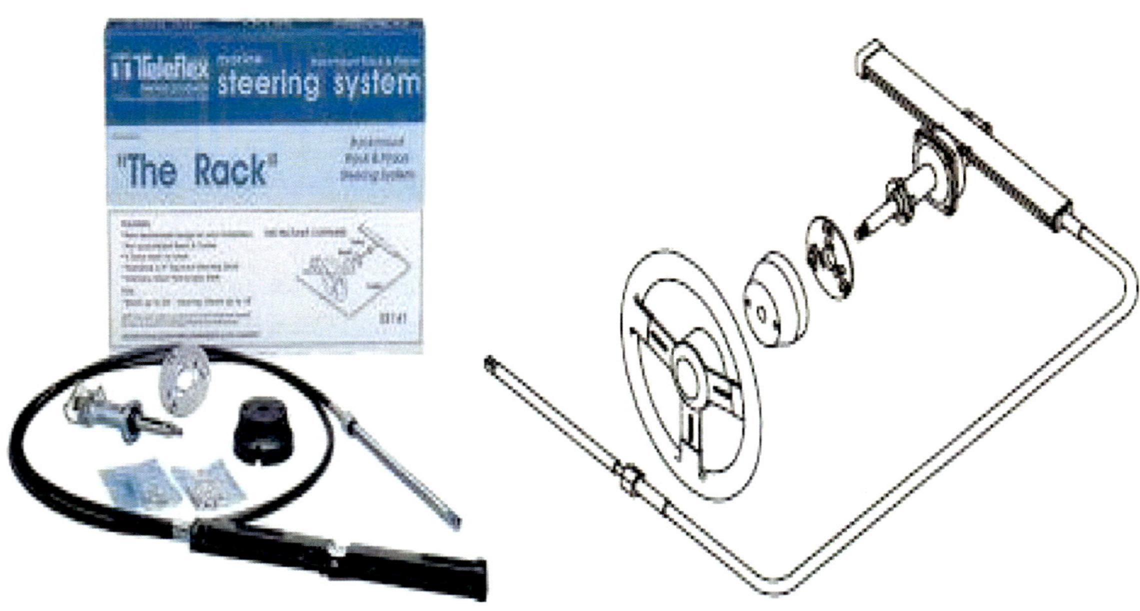

Mechanical Steering (Rotary)

Traditional mechanical steering is recommended for power assisted steering applications and some smaller inboards

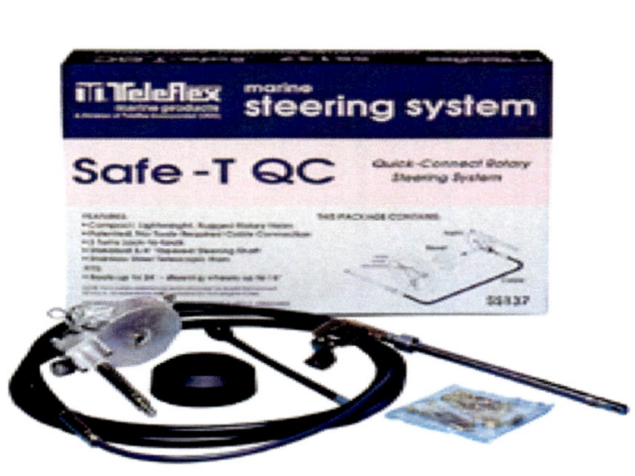

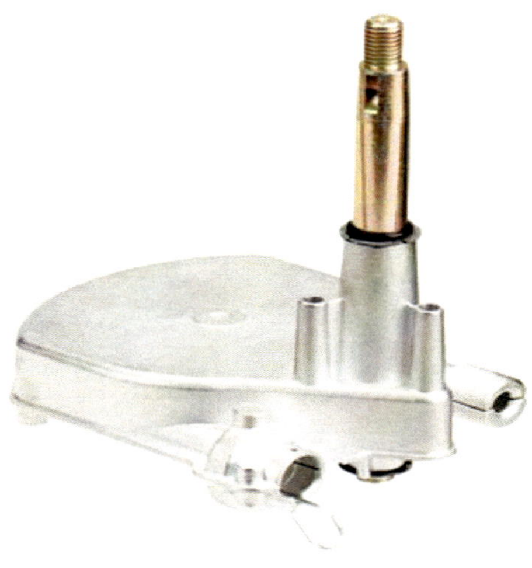

Rotary Steering Helm

Rotary steering uses a rotary gear in a housing mounted behind the helm station. The steering cable wraps around a drum driven by the internal gears inside the housing

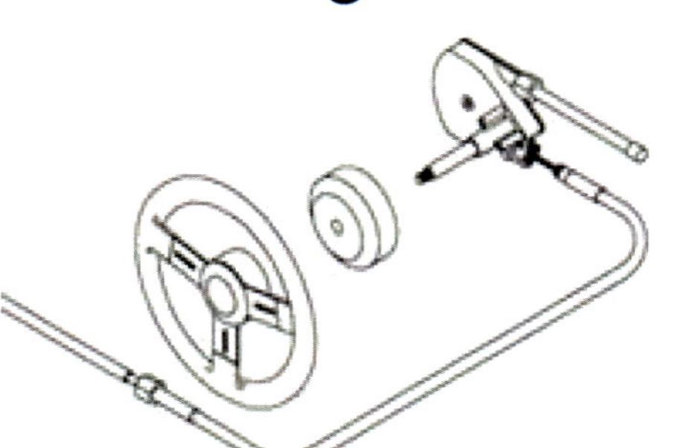

Rotary System

Safe-T was originally introduced in 1968. The original "The Rack" debuted in 1984, with the latest update (Backmount Rack) released in 1996. Millions were sold, resulting in a large replacement market.

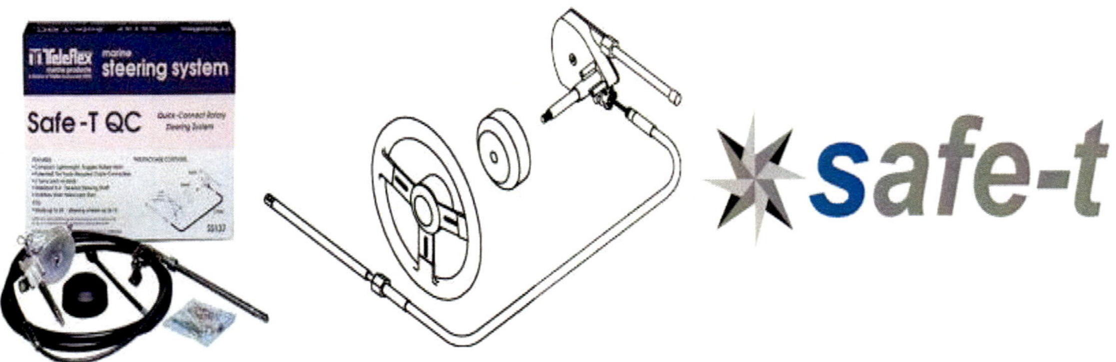

Rotary System

Safe-T QC and Backmount Rack helms are recommended for use with power-assisted steering and some small inboards. All systems meet ABYC requirements

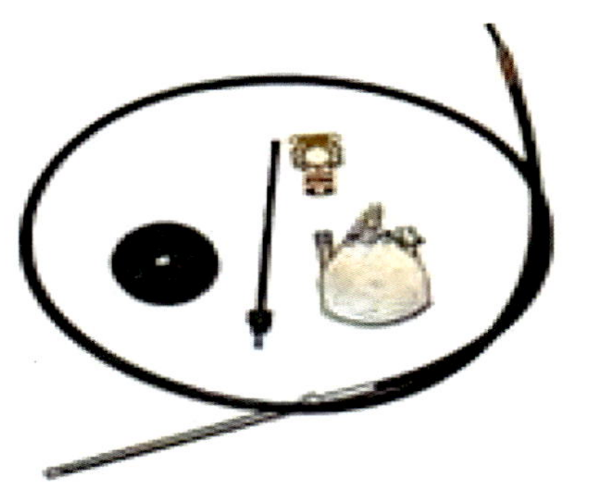 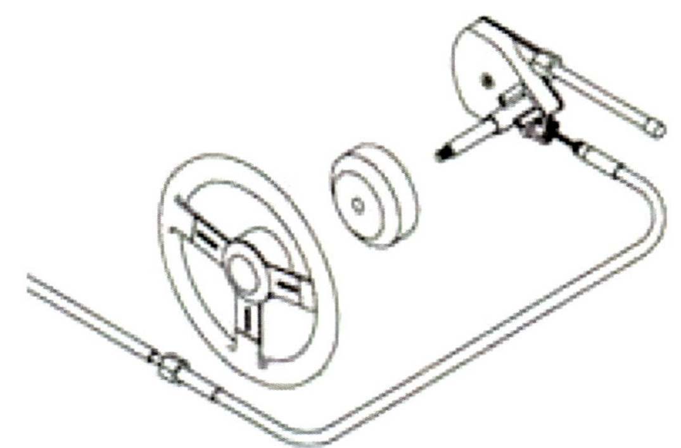

Mechanical Steering Troubleshooting

- Use a wrench to remove the nut and bolt at the link arm of your outboard

- Turn the steering wheel all the way to the left and all the way to the right, to ensure the wheel doesn't bind

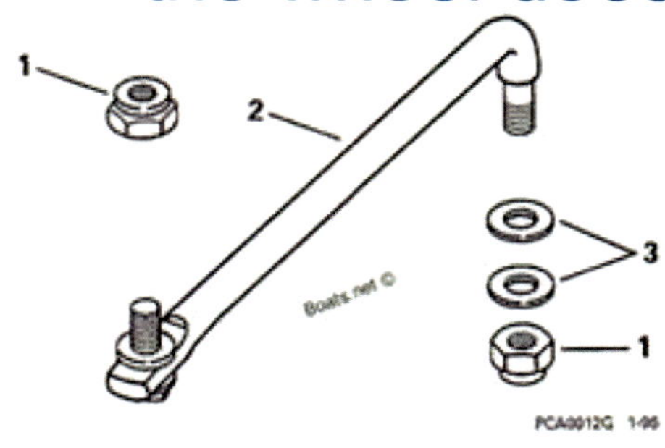

Mechanical Steering Troubleshooting

- If the wheel turns freely and a visual inspection reveals no deterioration, check the helm unit behind the wheel
- If there is deterioration and the wheel doesn't turn, replace the steering cable

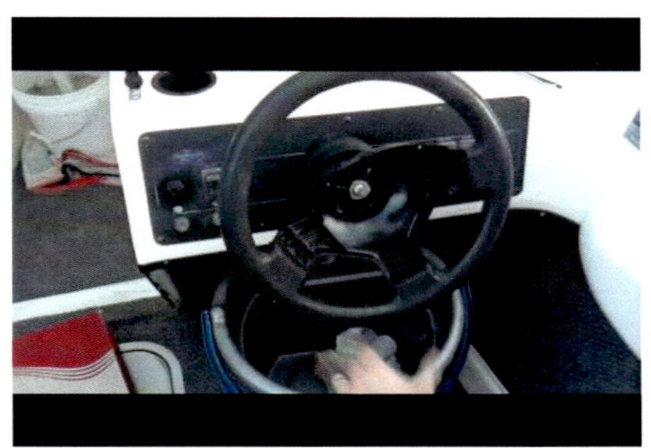

Mechanical Steering Troubleshooting

Use the wrench to disconnect the steering cable from the helm. Try to turn the wheel all the way in both directions .If the wheel doesn't move, the helm must be replaced

Mechanical Steering Troubleshooting

Move the outboard motor from side to side, though its full range of travel. If it doesn't move or moves only with great effort, the outboard should be lifted up and the bearings bracket shall be inspected to verify corrosion or moisture accumulation

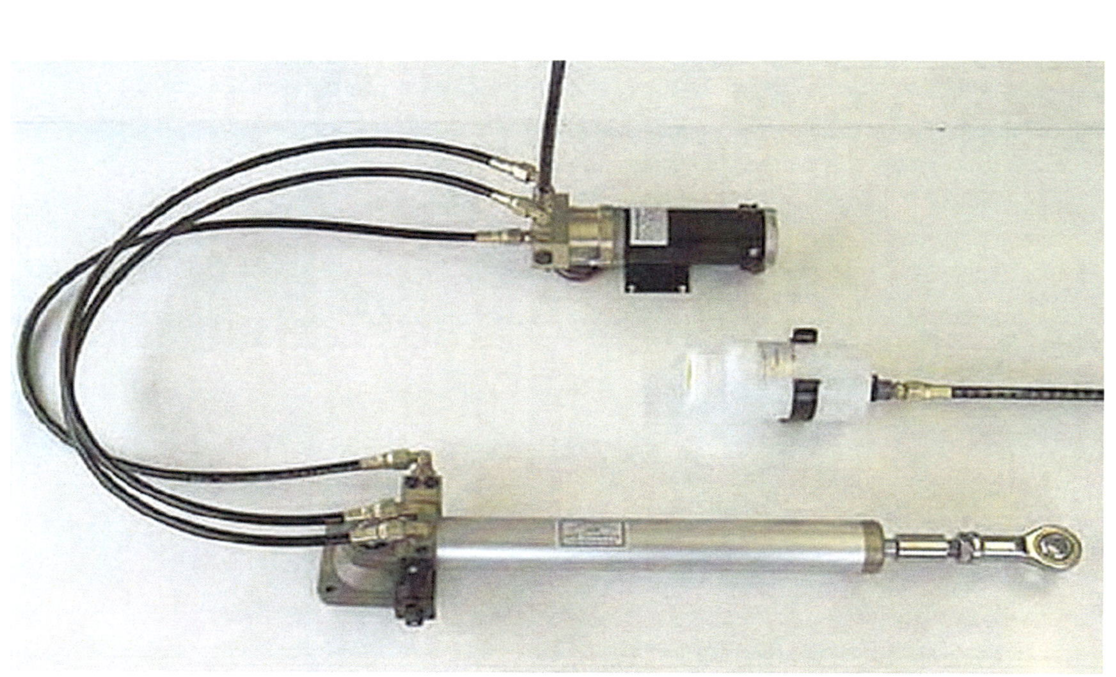

Power Steering Systems

Hydraulic & Hynautic

Boat Hydraulic Steering Systems

There are only two basic components in all the hydraulic steering systems. These are the helm unit and the cylinder, connected by nylon or copper tubing

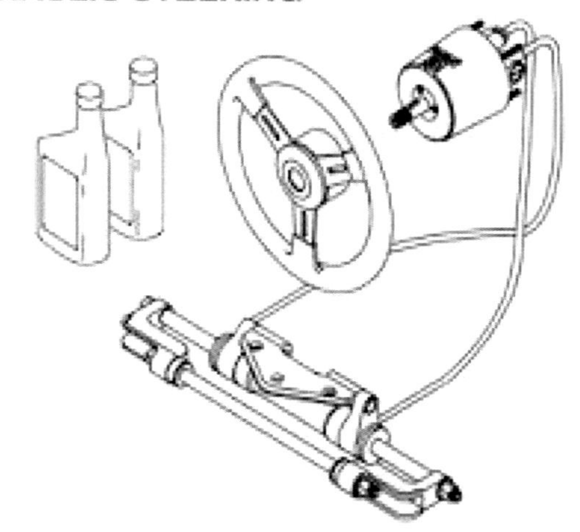

Teleflex Hydraulic Steering

- The SeaStar Hydraulic Steering System is designed to provide that extra margin of muscle when needed.

- The SeaStar System easily handles outboards, sterndrive and inboard boats.

- System is simplified with just three major components
    - Helm
    - Cylinders
    - Tube or Hose.

The Compact Helm

The helm unit consists of a pump section and a valve section, which includes an internal pilot check for multiple station operation

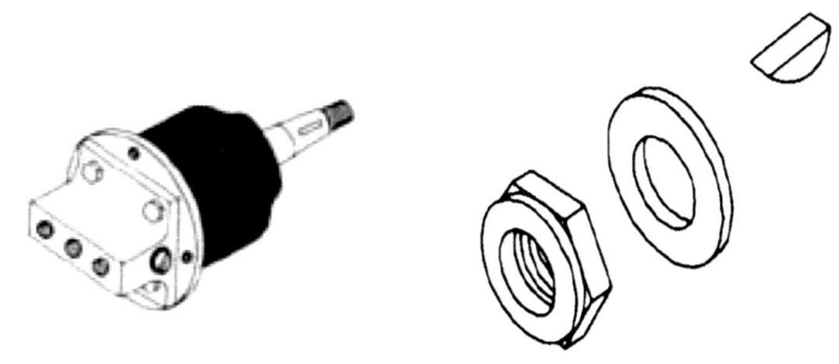

The Valve Assembly

The valve assembly prevents outgoing fluid from returning along the same line, isolates each steering station, locks the rudder and eliminates rudder "feedback" to the helm

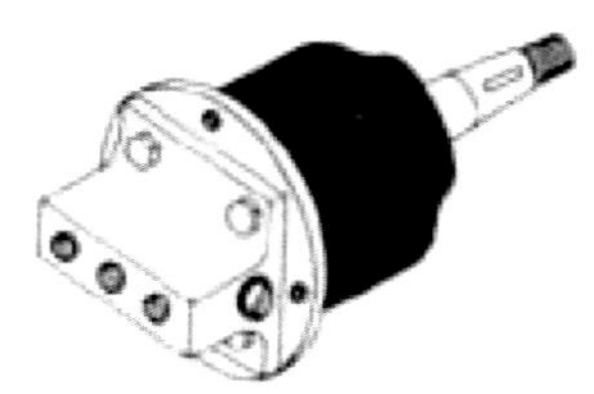

Typical Compact 2-Line

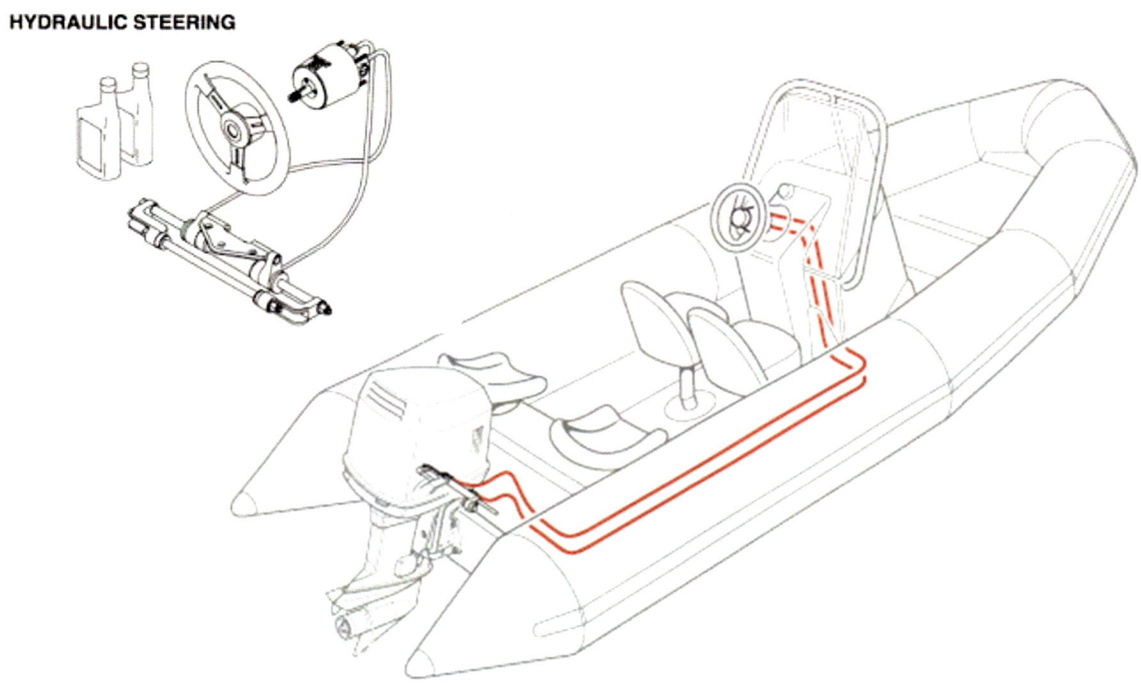

Compact 2 Lines or Compact 3 Lines

The difference between both systems is the addition of a third line to provide the return between the new reservoir tank and the new helm pump with three positions

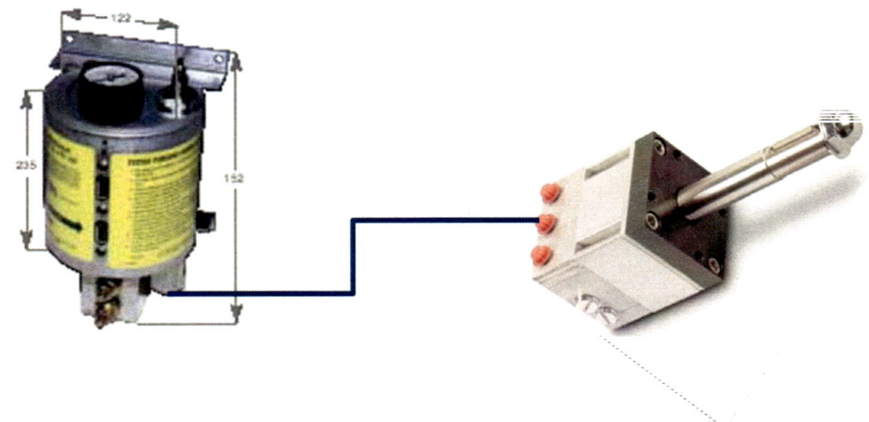

The Reservoir

The reservoir maintains the fluid reserve and pressure for the hydraulic system. Clear sight tube makes visual fluid inspection easy

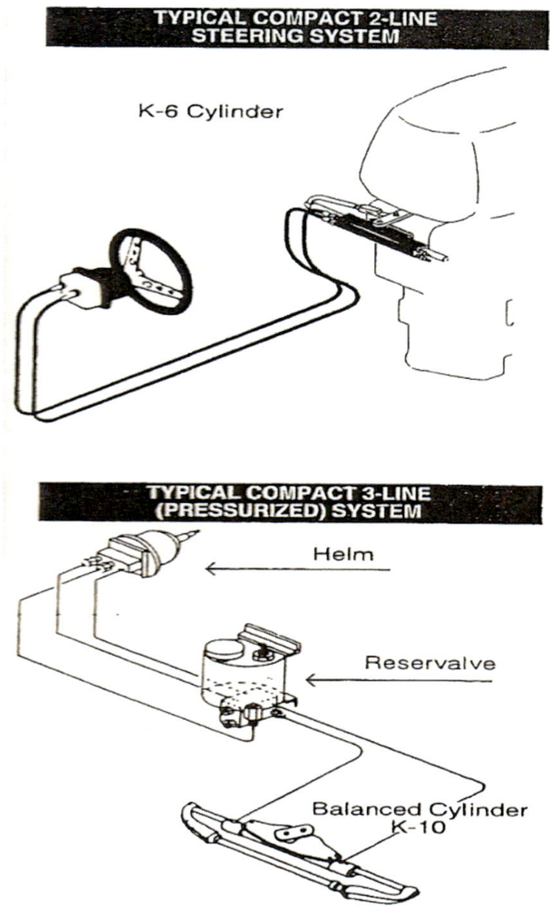

Inboard Hynautic Steering

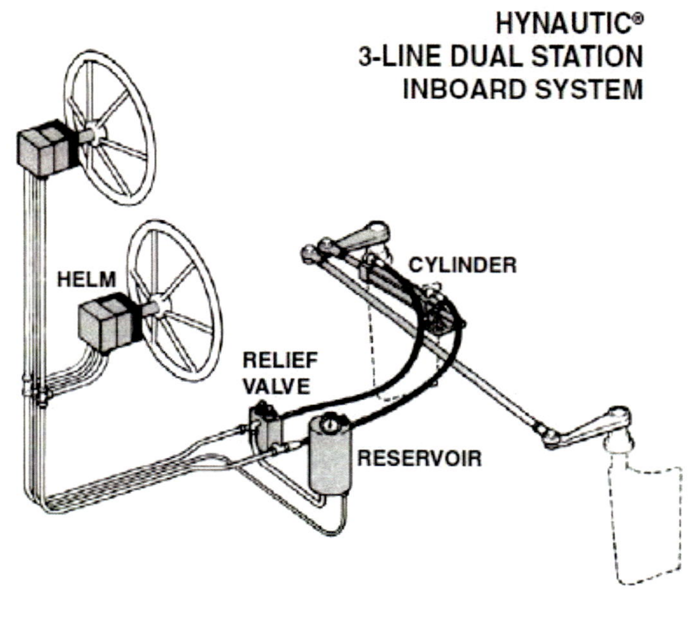

(turns vary by system)

SeaStar, BayStar and Hynautic

SeaStar, BayStar and Hynautic together have 99% of U.S. hydraulic steering market

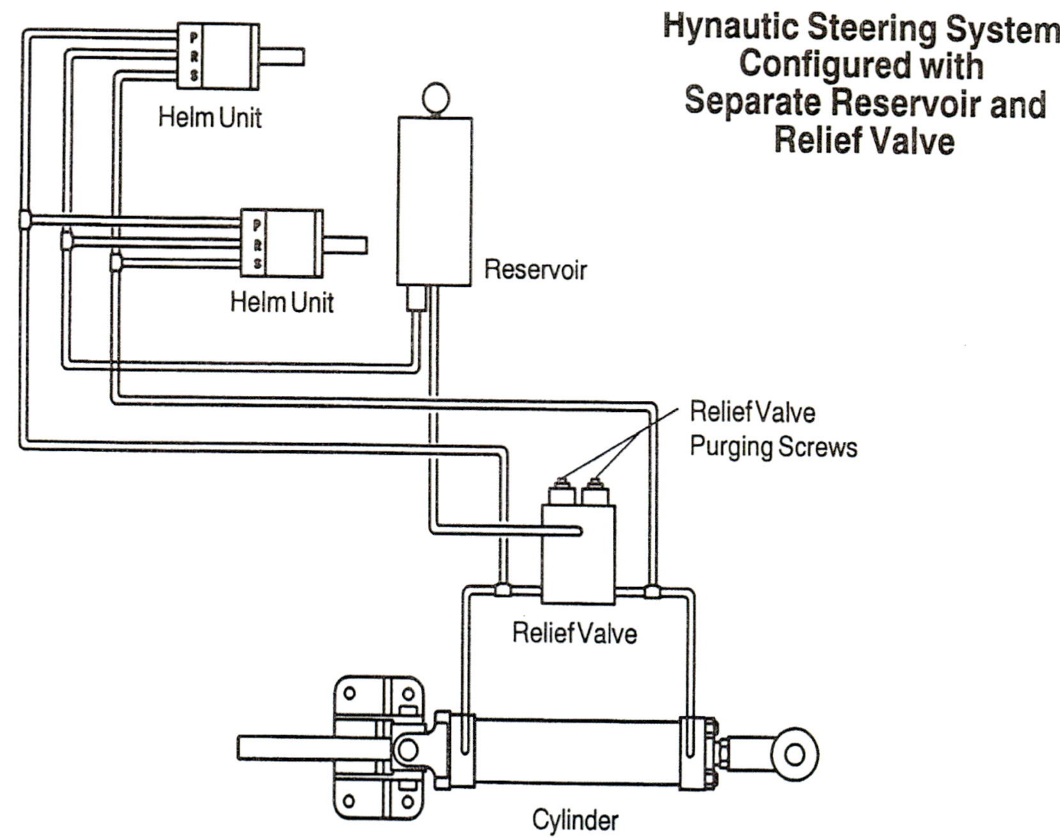

Power Assisted Steering System

The power assisted unit is an extra boost pressure produced by an electrical pump or by a mechanical pump usually driven by any engine crankshaft pulley

The Auto Pilot System

The power assisted unit is the heart of the Auto pilot system. It works together with the GPS , rudder sensor , the servo cylinder and the fluxgate compass

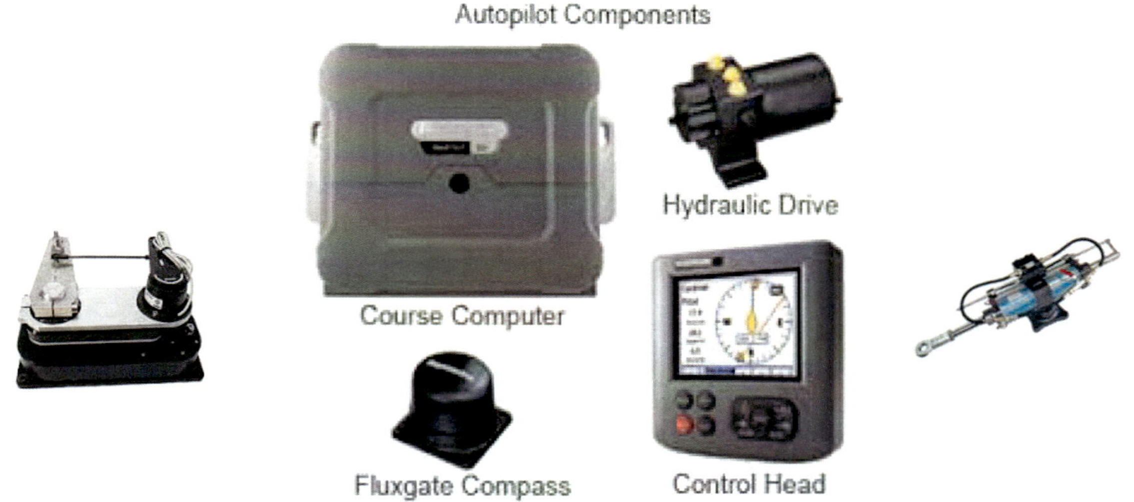

Power Steering Systems (Mechanical)

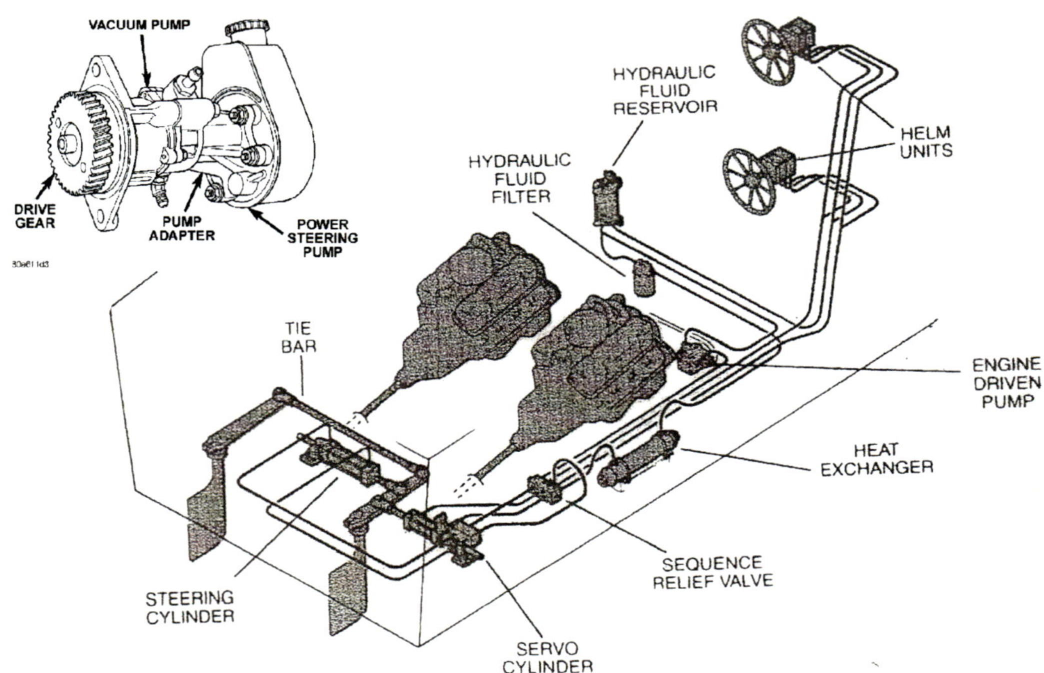

How Power Assisted Unit works

- This unit is an on-demand pump that only works once the helm pump send a hydraulic signal according with any movement of the helm

- If the power assisted unit receive a signal from the port side, then it sends a boost pressure over the same line between the power assisted unit and the servo cylinder

How Power Assisted Unit works

- The power assisted unit is installed in series between the helm pump and the Cylinder

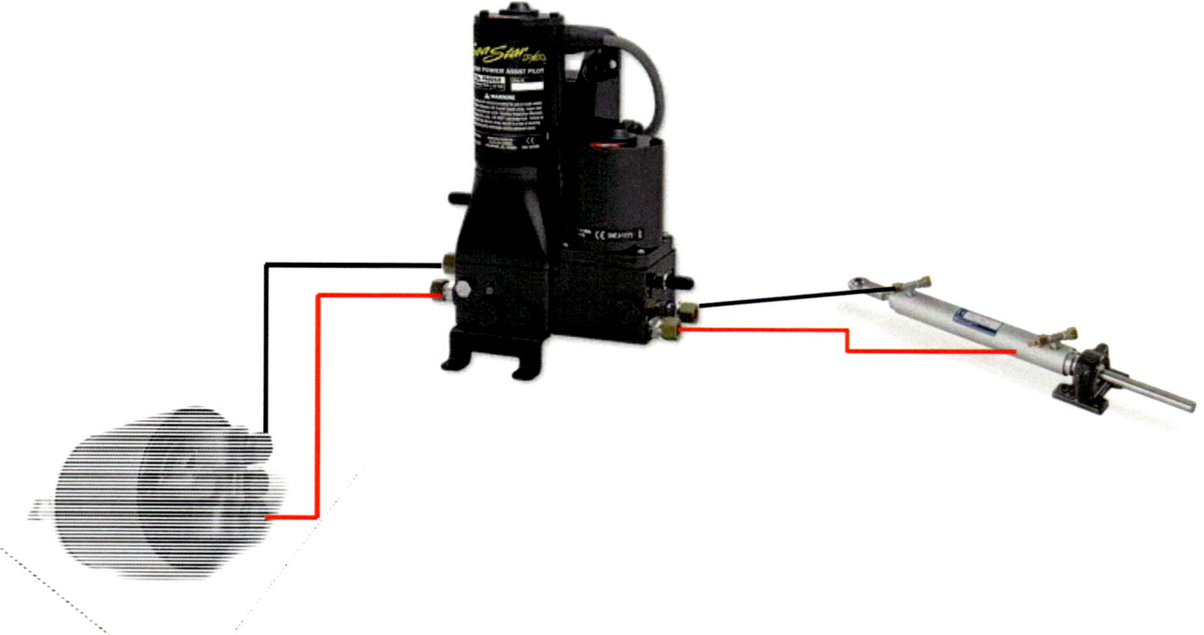

How the Power Unit Works with Auto-pilot

- The Servo Cylinder senses the current location of the rudder relative to the direction you want to steer and directs fluid from the Power Pump to the steering cylinder to alter course in that direction
- The fluid output from the pump is metered and controlled to ensure sufficient flow and smooth steering operation

How the Power Unit Works with Auto-Pilot

- When course correction has been achieved, the servo valve is neutralized and fluid is returned to the reservoir
- Circulating fluid from the power pump is cooled and kept clean with the heat exchanger and filter

Preventive Maintenance
(Every 6 Months)

- Check fluid level in the reservoir (Should be approximately ½ to ¾ of sight glass).
- Check System Pressure. It should be between 20 and 25 psi.
- Check for proper lubrication in pivot points and servo valves stroke guides.
- Check mounting bolts on the cylinder and servo to verify that vibration has not loosened them.

Preventive Maintenance
(Every 6 Months)

- Cylinder and servo ball joints must be kept free and well lubricated.
- Check clevis pins and/or ball joints.
- Check hoses for possible chafing or scuffing against one another or part of the vessel.
- Check belts in power pump for wear and proper tension.

Convert from Mechanical to Hydraulic

- If the boat is a single outboard you only need:
- -The Hydraulic Helm Unit

- -The Steering cylinder
- -The flexible lines

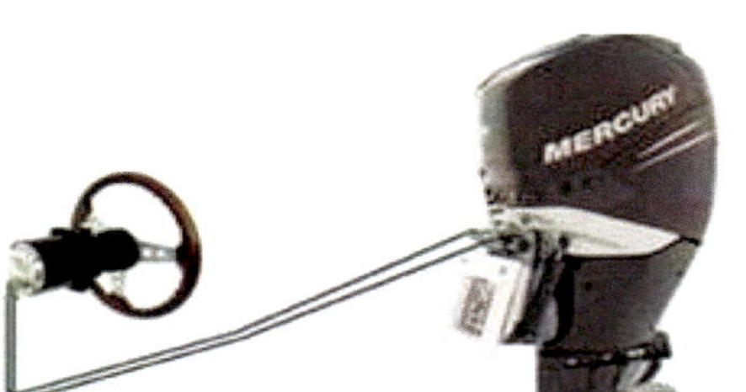

Convert from Mechanical to Hydraulic

- If the boat is a multiple outboard you need:
- -The Hydraulic Helm Unit

- -The Steering cylinder
- -The flexible lines
- -The reservoir tank
- -The Power assisted unit

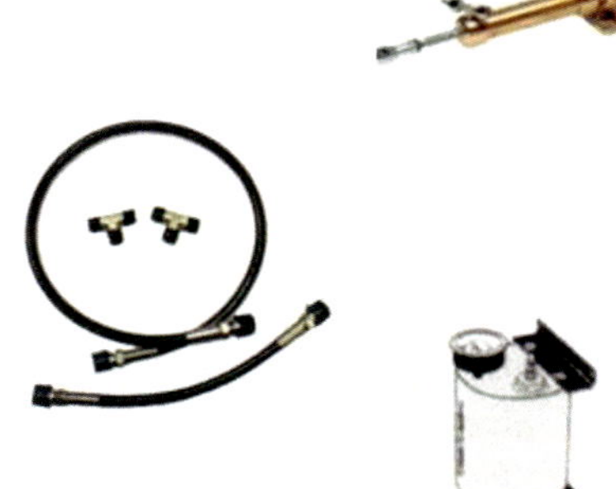

Chapter 6
Hydraulic Stabilizers
&
Side Thrusters

- How Stabilizers Works
- Principle of Operation
- Bow & Stern Thrusters
- Installation Procedure
- Hydraulic Bow Thrusters
- Hydraulic Motors and Pumps

Hydraulic Stabilizers

Normally in boats with more than 60 Ft length, trim tabs and stabilizers works together to keep the boat leveled

How Stabilizers Works

All stabilizers works on a principle of gyroscope. More specifically, the gyroscope tends to resist any power applied to it, but when it is applied it reacts so that its axes moves perpendicularly to the power applied

 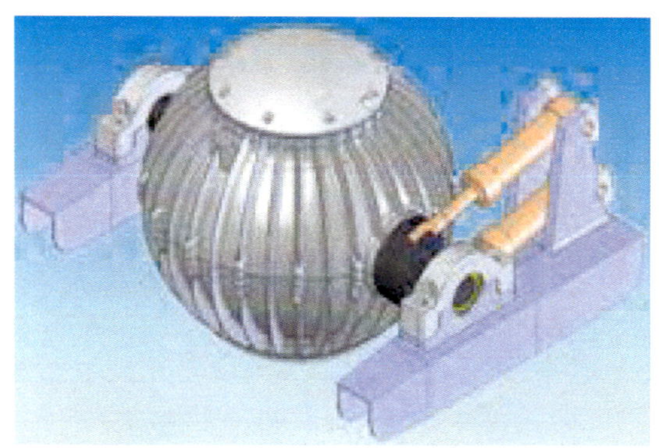

Hydraulic Stabilizers

- The whole staff, consists of a hydraulic plant, gyroscope and of coarse the fins
- Basically as the boat rolls the gyro axes moves fore and aft and so opens and closes the appropriate valves.
- The only thing which can be controlled by the user, is synchronization of the movement of fins with rolling of the boat

Hydraulic Stabilizers

- All stabilizers are effective while rolling, they have no effect on pitching of the boat

- Tough there are no specific limitations on the size of the boat where they should be installed, it make sense to install them on boats 60 feet and over

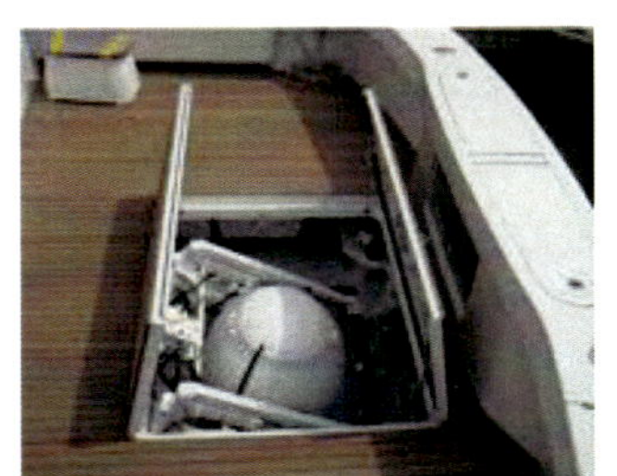

Hydraulic Stabilizers

New stabilizer systems include roll, pitch and yaw (steerage assist in following seas) Control. The most advanced system uses digital control and dual T-Foils (from hydrofoil technology) mounted on the transom. It controls all motion underway and roll at anchor

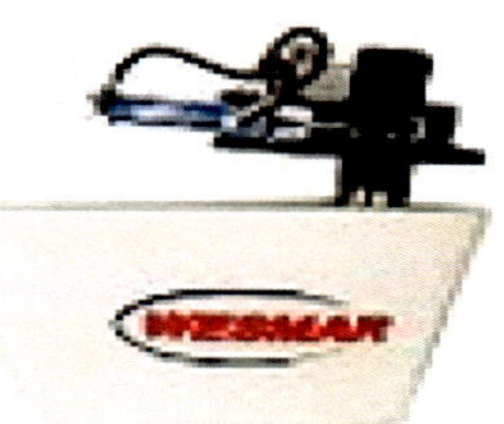

Stabilizer Hydraulic Systems

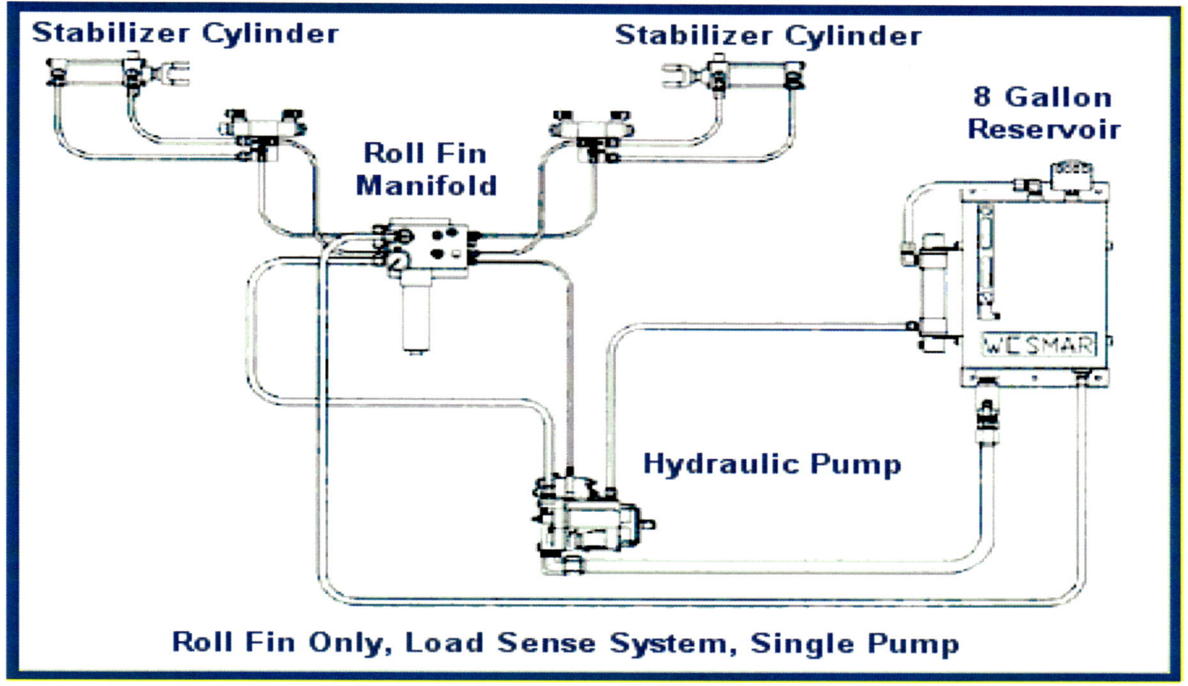

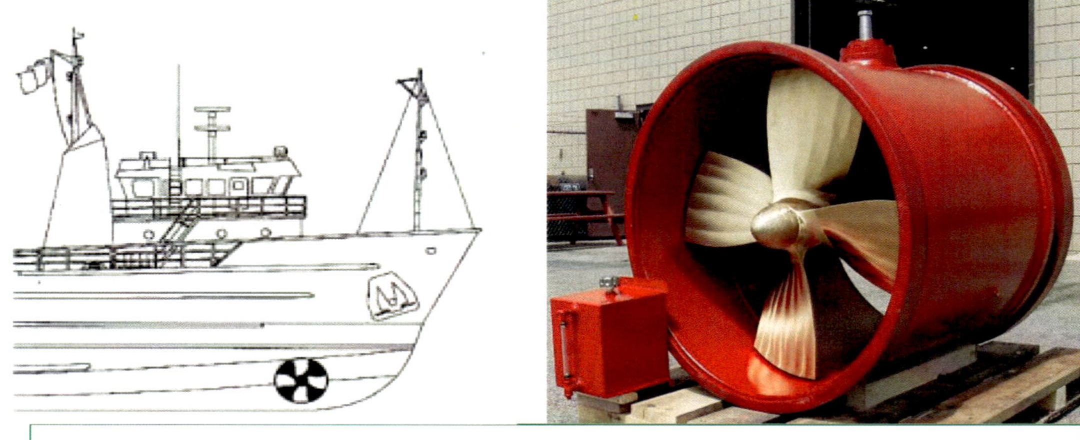

Bow & Stern thruster

- Installation Procedure
- Electrical Wiring
- Hydraulic Bow Thrusters

Bow & Stern Thruster

- A **bow thruster** is a transversal propulsion device built into, or mounted to, the bow of a ship or boat to make it more maneuverable
- Bow thrusters make docking easier, since they allow the captain to turn the vessel to port or starboard without using the main propulsion mechanism which requires some forward motion for turning

Bow & Stern Thruster

- An impeller in the tunnel can create thrust in either direction which makes the ship turn
- Most tunnel thrusters are driven by electric motors, but some are hydraulically powered
- Ships equipped with tunnel thrusters typically have a sign above the waterline over each thruster on both sides, a big white cross in a red circle

Taking the Boat out

The first step is move out of the water the boat and support it properly leveled

Leveling Boat

Seeking the best location

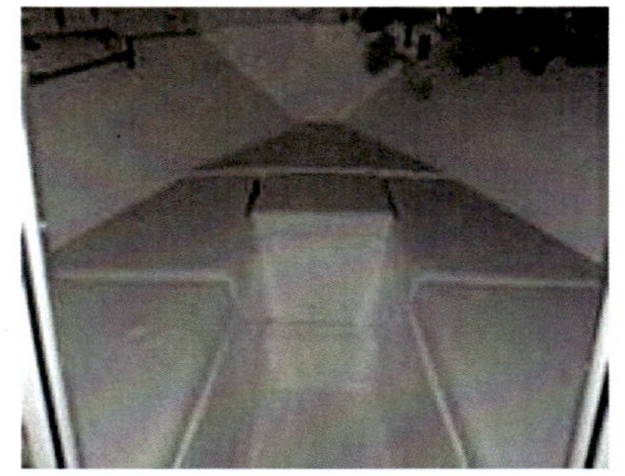

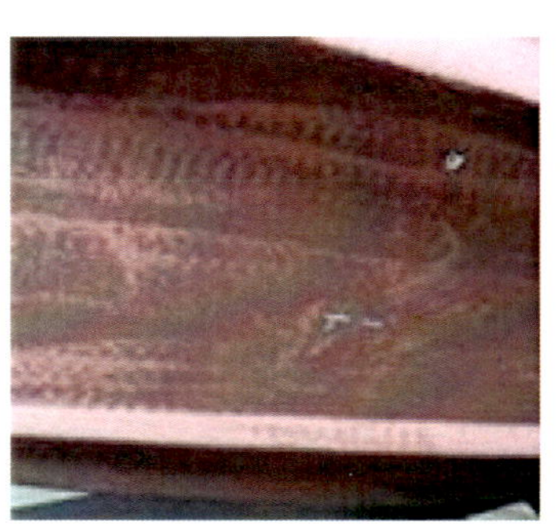

Identifying the Water Line

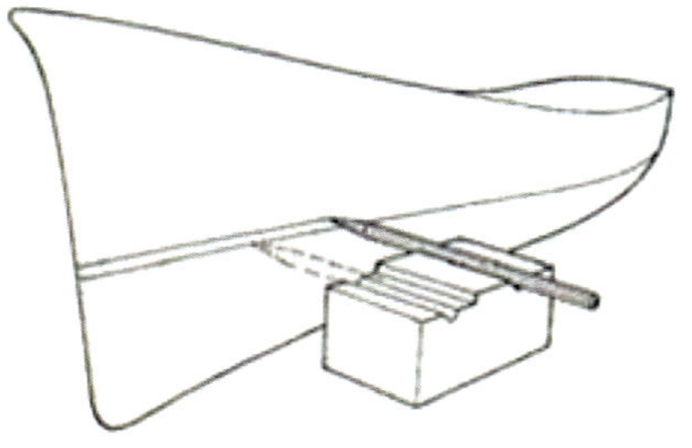

Marking the Hole for the Tunnel

Drilling the Pilot Hole

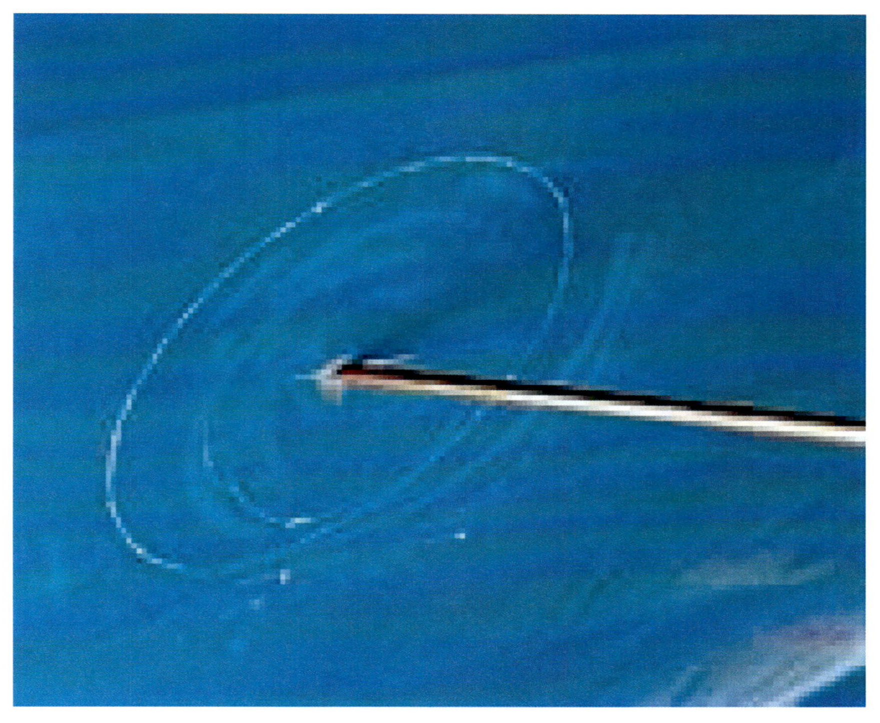

Marking the Hole for the Tunnel

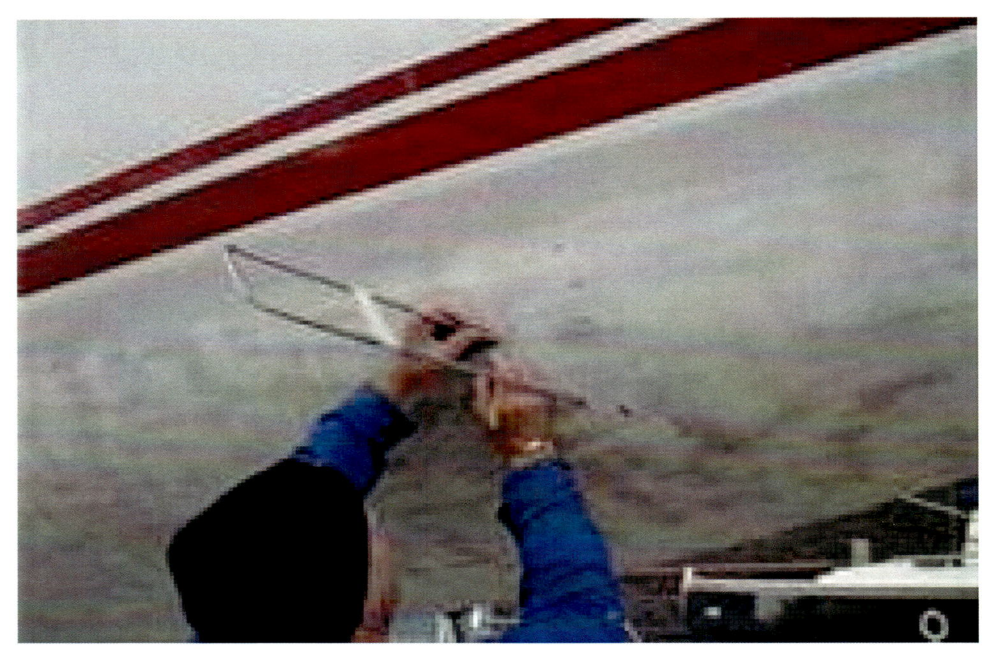

Bow thruster Tunnel

Select the tunnel pipe according with the propeller diameter and the hull material

Drilling the Hole

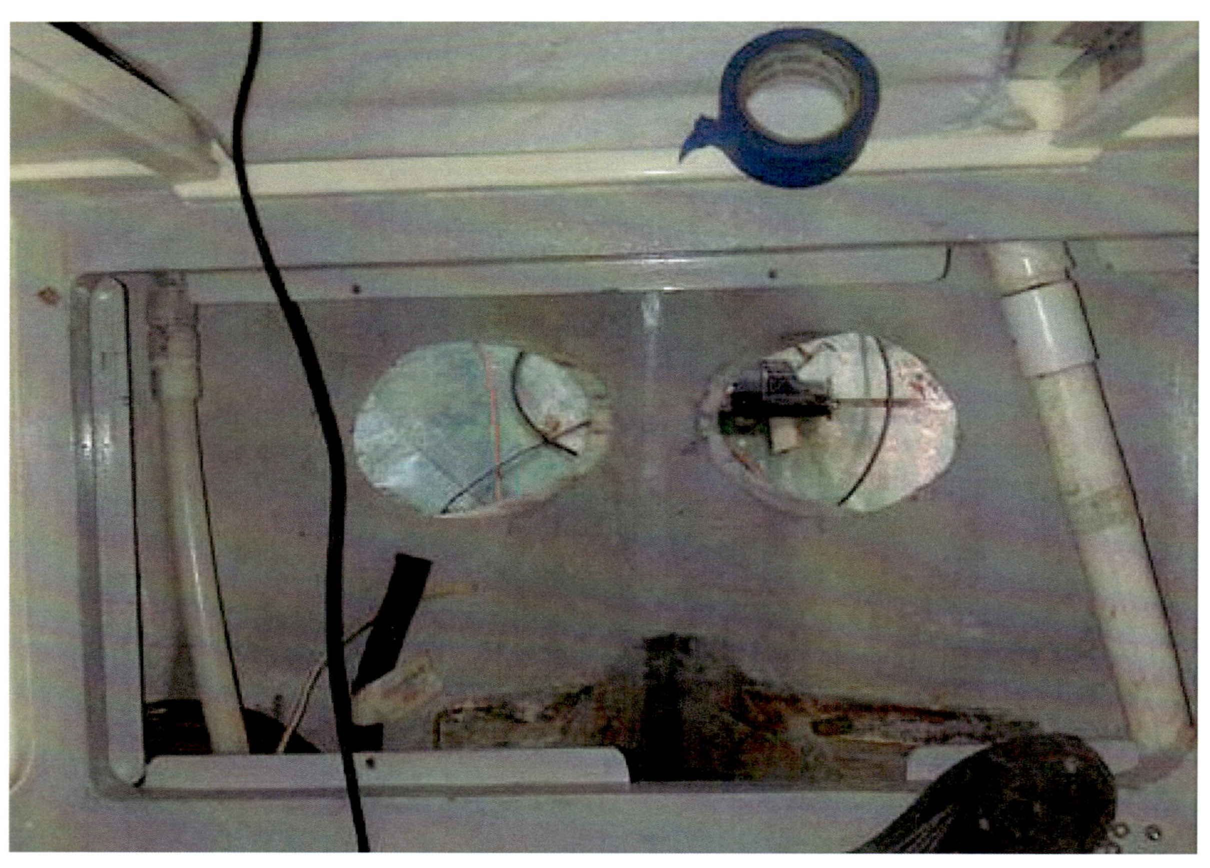

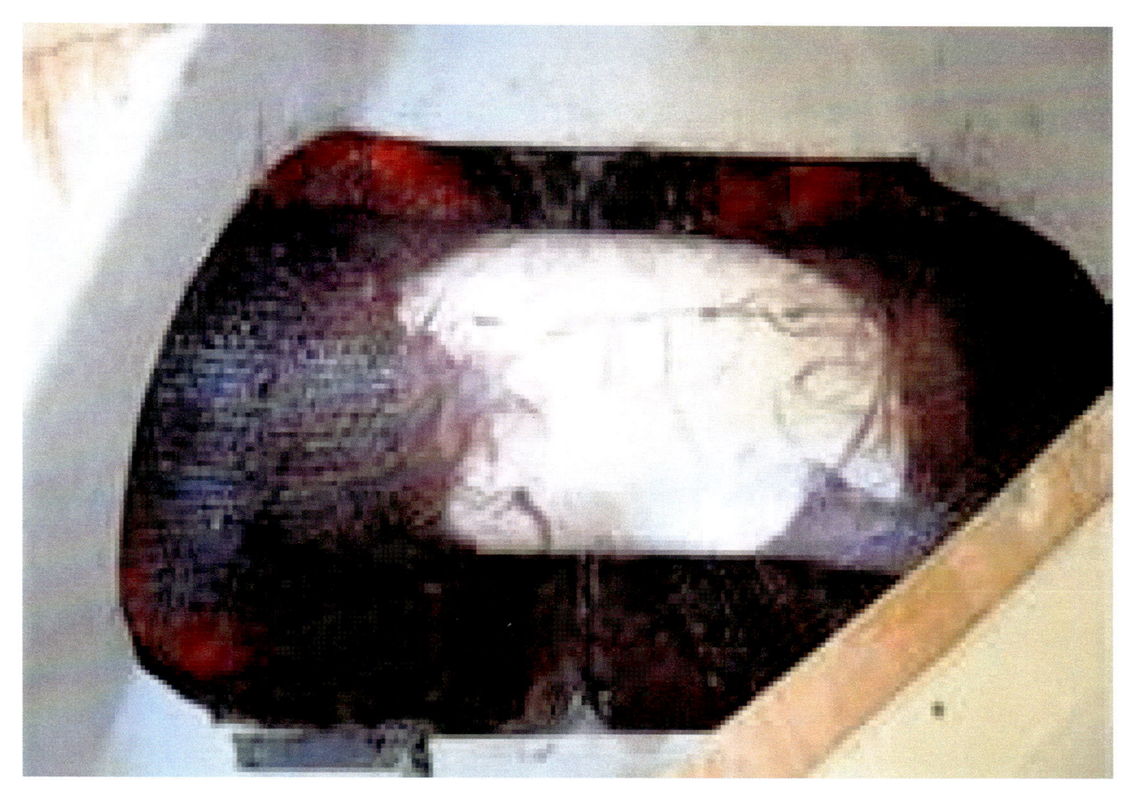

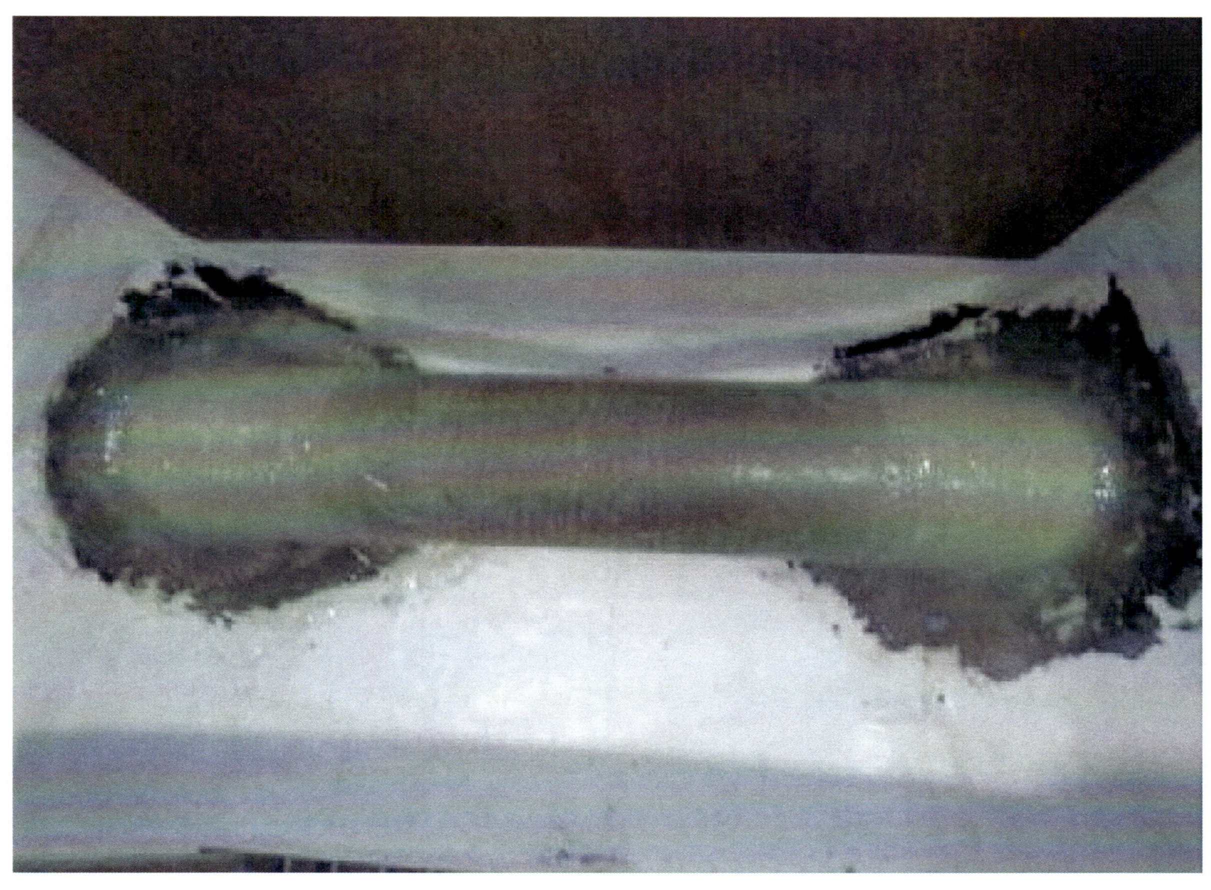

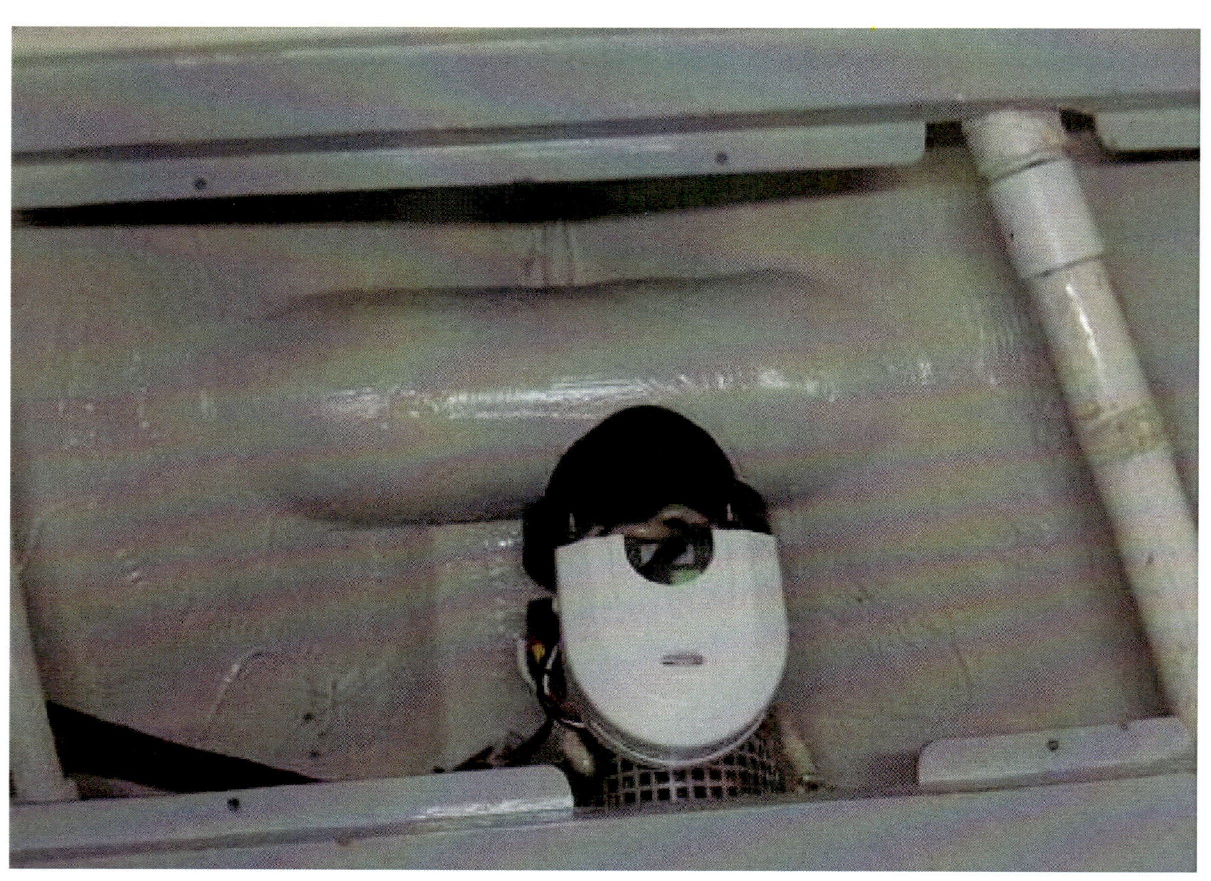

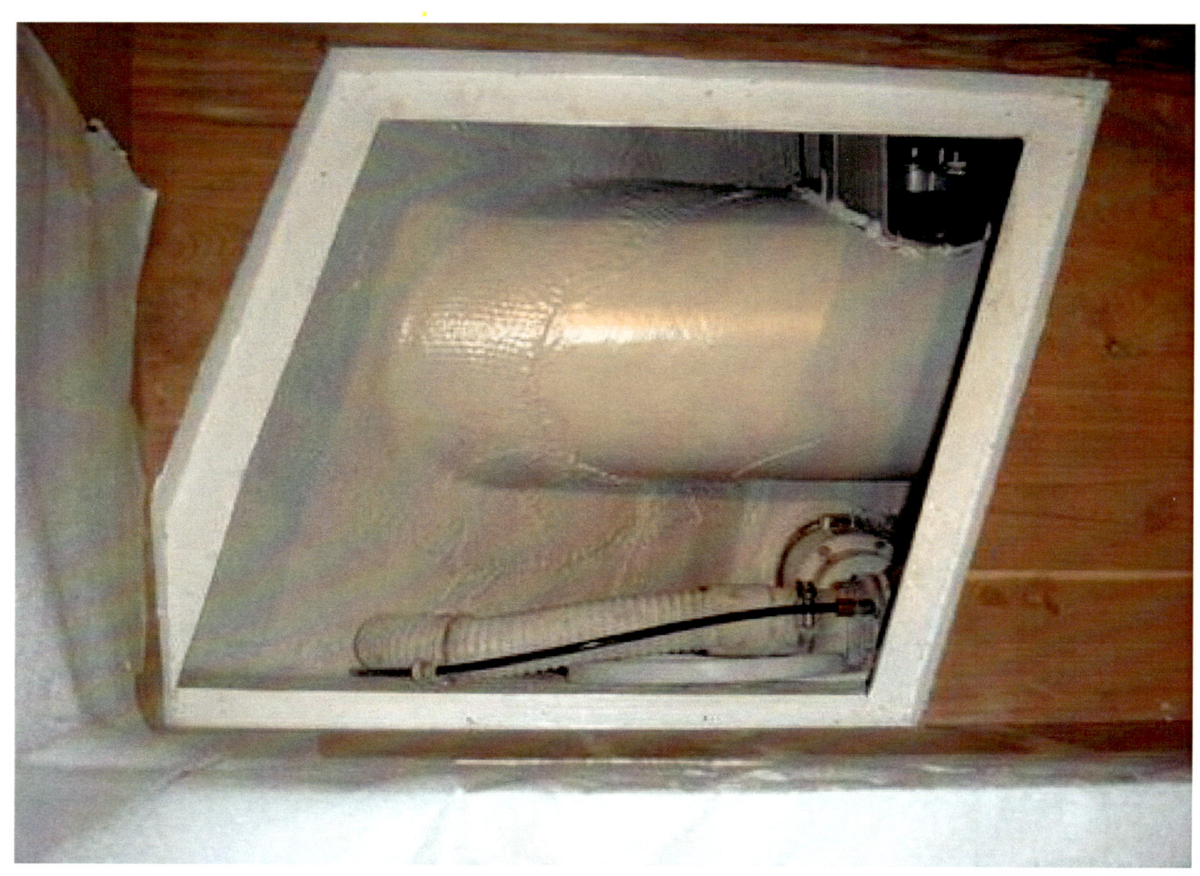

DC Motor Wiring

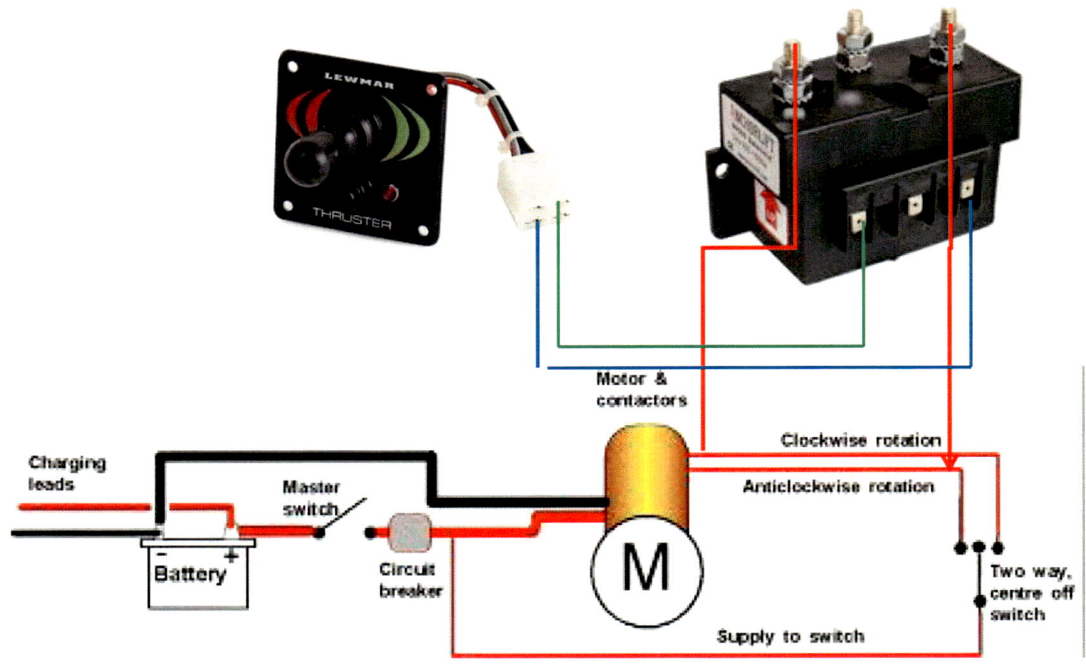

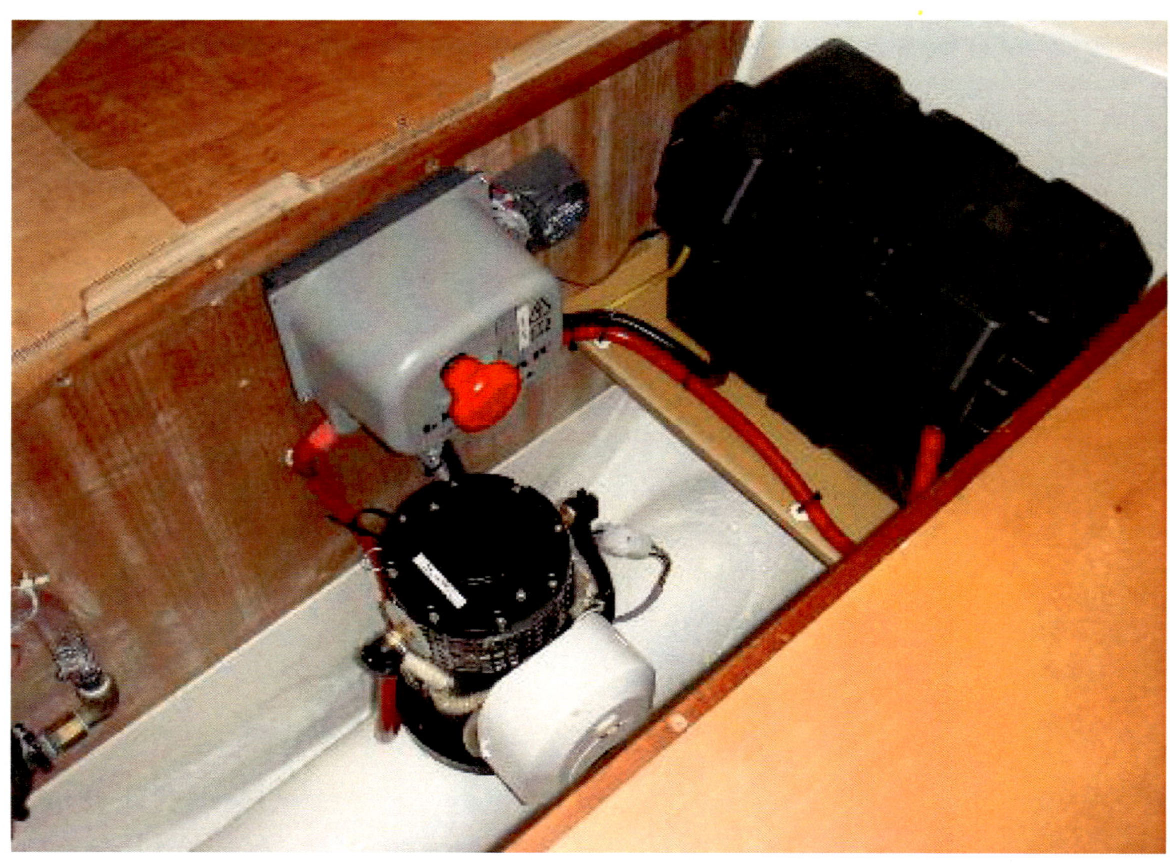

Hydraulic Bow Thruster

The hydraulic motor is driven by a hydraulic pump.
If a hydraulic pump and associated tank are
already installed on board, then , in most cases,
this assembly can be used for driving the bow
thruster too

*The "Invisible Man"
on board.*

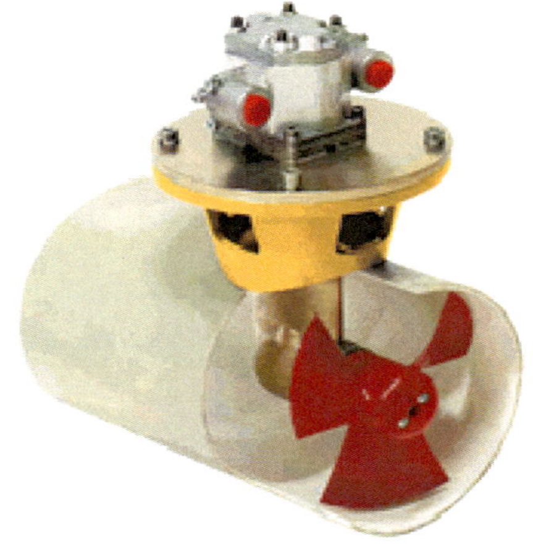

Hydraulic Bow Thrusters

Hydraulic Bow Thrusters

- Load Sensing System is More Efficient
- Doesn't Require Over-Speed Control
- Powerful, Quiet and Reliable
- Full Thrust Available at Engine Idle Speeds

Hydraulic Motors & Pumps

- A **hydraulic motor** is a mechanical actuator that converts hydraulic pressure and flow into torque and angular displacement (rotation).

- A hydraulic pump uses mechanical energy to create fluid flow. It draws hydraulic fluid from a reservoir and pushes it into a hydraulic system which is used to power

Electric Power Vs. Hydraulic

- Conceptually, a hydraulic motor should be interchangeable with a hydraulic pump because it performs the opposite function - much as the conceptual **DC start motor** is interchangeable with a **DC alternator**.

Hydraulic Motors

- The function of hydraulic motors is to convert hydraulic pressure and flow into rotational mechanical energy via an output shaft.

- Motors are used where a rotational output is required and actuators are used where linear output is required.

Output Power

- The output of a motor is a torque and an angular velocity or RPM.
    - Power = Torque x angular velocity
    - P = T X RPM.

- The hydraulic motor of the bow thruster is powered by the high pressured fluid from the hydraulic pump driven by the diesel generator

Hydraulic Motors

- The output torque of the hydraulic motor is around 90% of the torque produced by the diesel engine

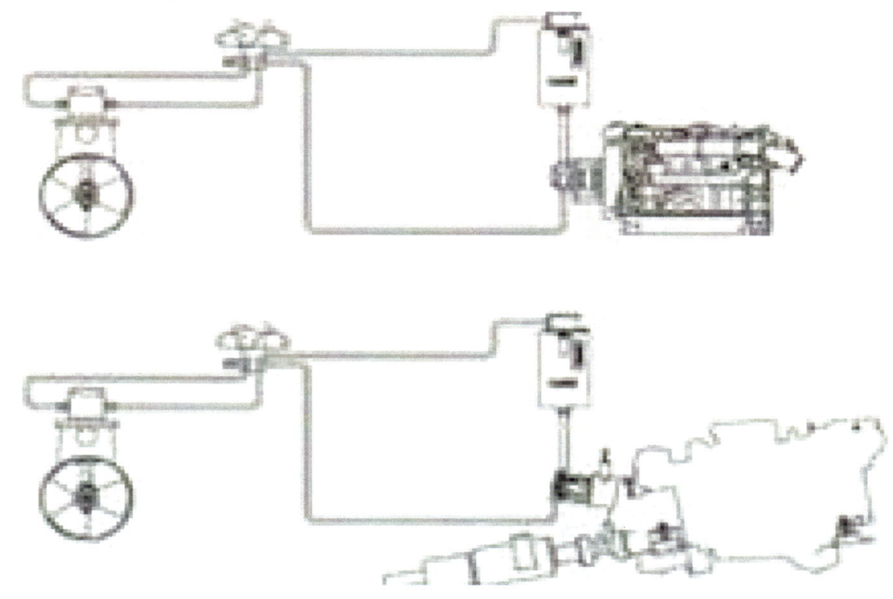

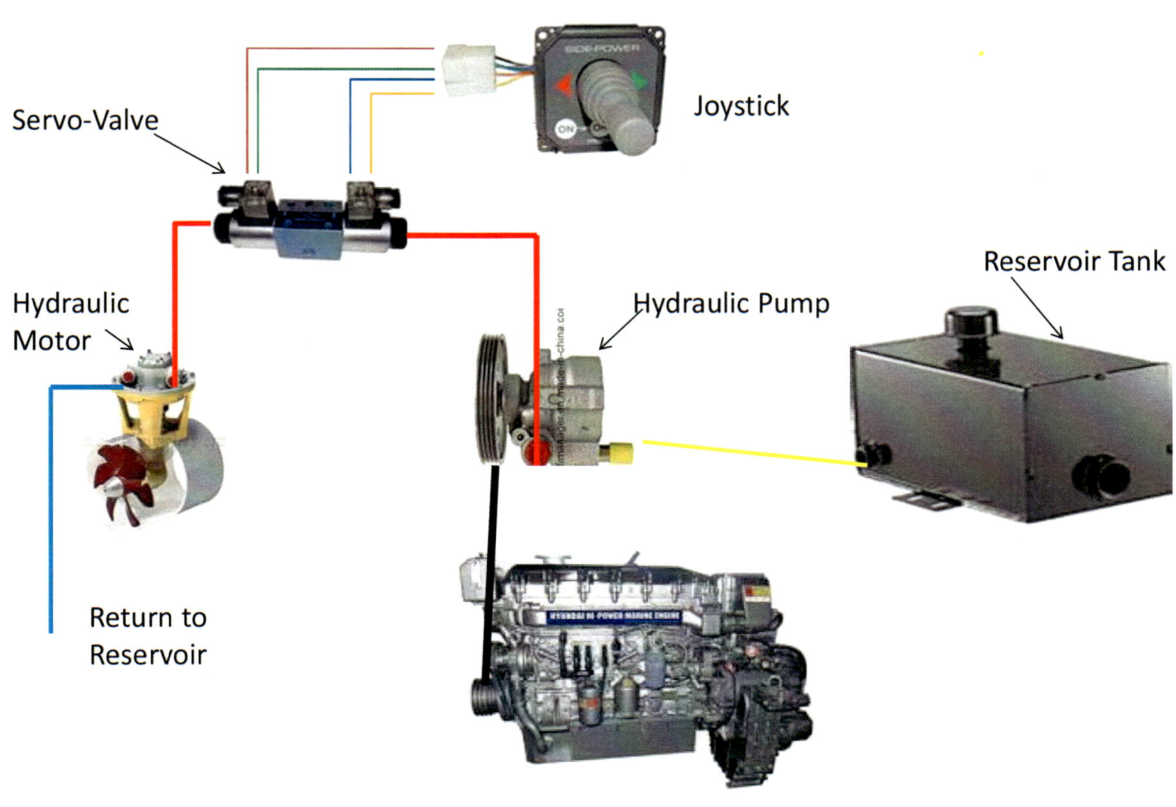

Hydraulic Bow-thruster Schematic

Chapter 7
Fresh Water Systems

- Fresh Water Systems
- Fresh Water Pumps
- Water Pumps Calculations
- Water pumps Configurations
- Pipes and Fittings
- Water Pressure Switch & Valves
- Water Tanks & Water Heaters
- Faucet and Shower
- Sump Pumps

Fresh Water Systems

Fresh Water Installation

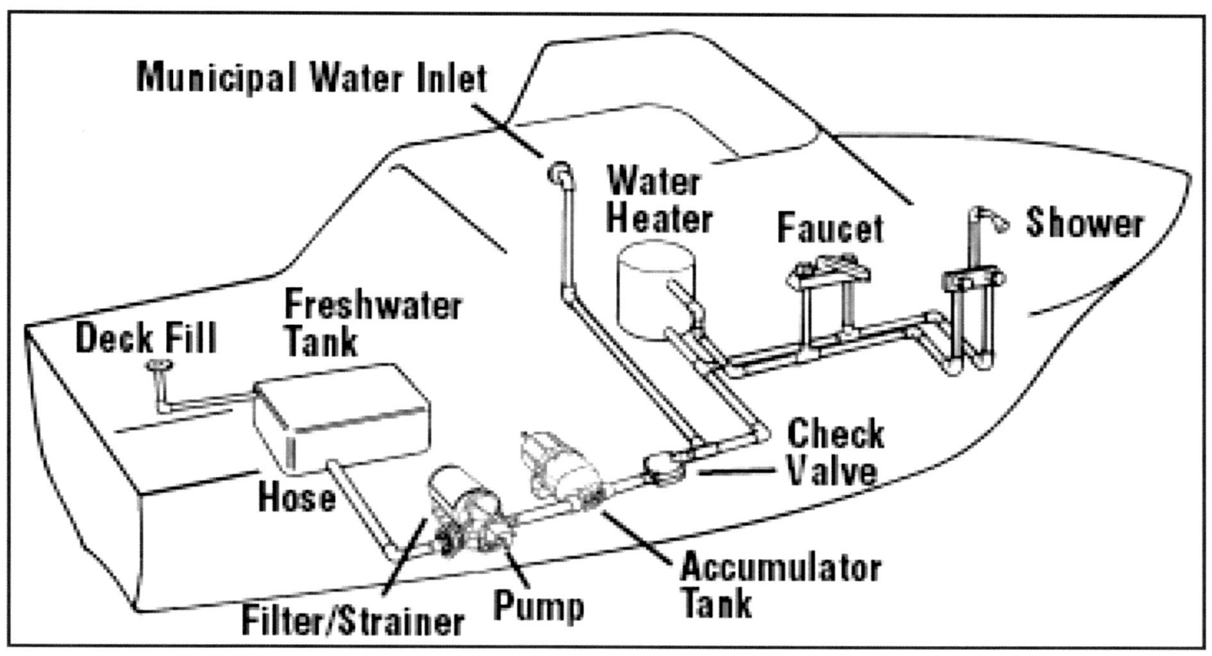

Municipal Water Inlet

- By connecting a drinking-water-safe garden hose between a municipal water inlet and a faucet on shore, you'll have a continuous source of pressurized water at the dock without ever having to fill the tank or operate the pump

- To protect and isolate your pump and accumulator tank from potentially damaging high pressure, install a one-way check valve as shown in the diagram

Fresh Water Pumps

- Pressurized water systems make life aboard more comfortable by providing water "on tap" for dishwashing, showers and other applications.
- The complexity of installing and maintaining one will depend on the number of outlets and accessories you choose.

Where they are used?

- On all boats larger than runabouts and day-sailers, fresh water on board is both a convenience and a necessity
- The freshwater pump is at the heart of the delivery system that ensures a constant supply to the fixtures in galley, head and shower

Demand /Delivery Pumps

- On-demand Pumps are ideal for applications that require high pressure with flow 1 GPM to 6 GPM and low amp draw
- They can be mounted in any position, are compact, and are designed for long, trouble-free life

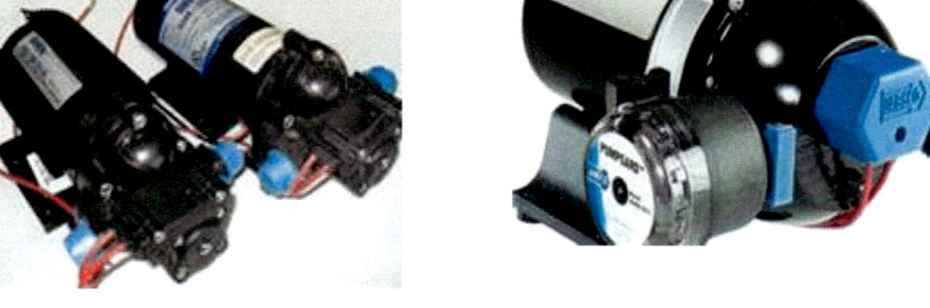

How many Outlets does the Pump need to serve?

- Fresh water pumps are frequently described by how many outlets they can supply ("for 2-3 faucets")
- Additional taps or fixtures in the system might cause a reduction in flow capacity, erratic flow, or excessive pump cycling if several were "open" at one time

How Much capacity do you want?

- Pumps will generally be selected by how many gallons per minute (GPM) they can pump, and/or by how much pressure they can develop (PSI).

- The freshwater pumps we offer have capacities between 1.1 and 11 gallons per minute. Some general guidelines for sizing the pump

Number of fixtures	Recommended GPH
1-2	Up to 3.0 GPH
2-3	Up to 4.0 GPH
4+	More than 4.0 GPH

Fresh or Salt Water?

- Many galley pumps cannot be used with salt water, because salt will damage the valves and seals or corrode internal metals

- However, to conserve fresh water, especially on small vessels with limited tankage, you might want to use salt water for dishwashing

- A popular method uses a bucket for the first wash cycle before rinsing the clean dishes with fresh water, either in the galley sink or with the deck shower system.

Where Do you Mount the Pump

- This question determines whether you need a self- or non-self-priming pump
- Non self-priming pumps have to be mounted at or below the water level, so water is already in the pump chamber. They can push water uphill against gravity

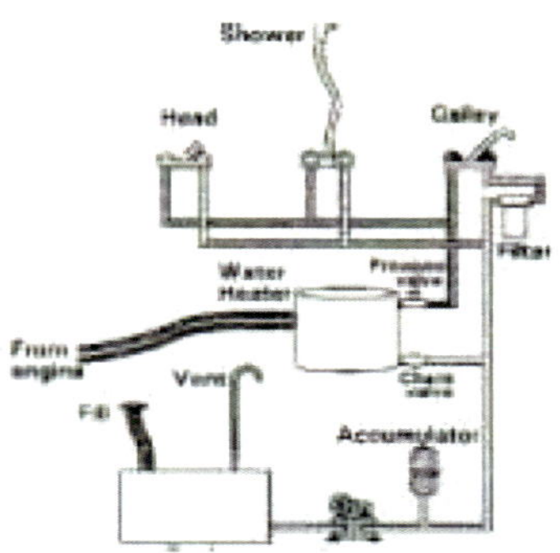

Self Priming Pumps

- **Shurflo** Freshwater System Pumps for Boat & RV applications are self-priming and employ three independent pumping chambers to pump water smoothly and quietly. Built-in check valves prevent backflow into the storage tank. They can even run dry without damage. Thermally and Ignition Protected.

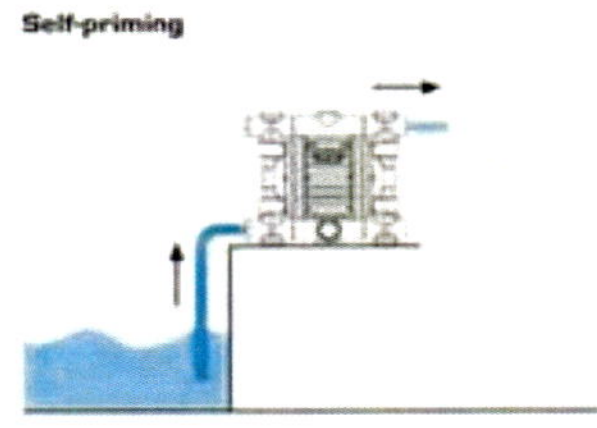

Which port size do you need?

- Higher capacity pumps will have port sizes of 3/4 or 1" to allow more flow
- Some shower pumps offer multiple port sizes between 3/4" and 1 1/2". Otherwise 1/2" NPT ports are very common on freshwater pumps

Do you want on-demand water supply?

- On-demand pumps have a pressure switch that makes the pump build up pressure in the fresh water line
- Whenever water is demanded and the faucet is opened, a stream of pressurized water will be available. The pump turns on to maintain this pressure as needed

Adjustable on-demand water supply

- You may also choose to turn up the flow and enjoy a high-pressured shower for "Just Like Home" comfort, even while another fixture is simultaneously in operation

- With the Smart Sensor™ 4.0 and 5.7 the power of flow is in your hands. Delivering flows from 4.0 GPM up to 5.7 GPM, these 12 VDC pumps are also available in 24 VDC

Adjustable on-demand water supply

The Smart Sensor™ 4.0 is designed specifically for cruisers and midsize yachts. Delivering 4.0 GPM and pressure up to 50 PSI, this micro-controller based variable speed pump precisely monitors your system's water pressure and adjusts the motor speed, eliminating the need for a pressure switch

On-demand or Constant Pressure

By contrast, manual pumps (Constant pressure) have no pressure switch, and must be turned on manually (via an electric switch) prior to when the water is required

Variable speed pumps

- New variable speed pumps offer a noticeable improvement **over accumulator tank systems**.
- They change the speed of the motor to deliver the precise amount of water wherever it is needed
- There's virtually no lag time, and there's no drastic difference in pressure if two or more faucets are open simultaneously

Rated GPM

- Gallons per minute. Describes the pump's output under ideal (open flow) conditions
- This does not take into account head height, friction in the system and other factors that will reduce output
- For most boats, high GPM pumps may not be an advantage, especially if the fixtures only permit a limited amount of water to flow.

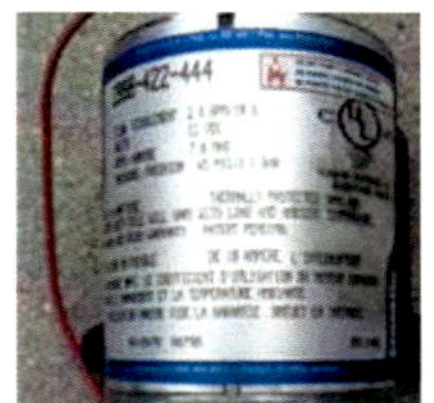

PSI or Head ??

- Pump users want pressure. Pump manufacturers supply feet (or meters) of head
- In the final analysis, they are the same, just expressed from two different points of view
- As someone who specifies and/or installs pumps, you need to know how these terms relate to each other

Convert PSI into Head

- In simple terms, the mathematical constant 2.31 converts a unit of energy against gravity into a unit of force against any other area
- This constant converts a foot of head of water into pressure: Head in feet of water divided by 2.31 equals pressure in psi, and pressure in psi times 2.31 equals head in feet

- The pressure - *psi* - of a water pump operating with head *120 ft* can be expressed as:

- *p = (120 ft) 1 / 2.31   = <u>52</u> psi*

Convert PSI into Head

- If I want to buy a pump that develops 30 psi to pump water, what is my pump rating? What pump should I buy?
- 30 psi x 2.31 = 70 feet
    - If you need a pump to develop 30 psi of water pressure, then buy a pump that develops 70 feet of head.

Where does the Constant 2.31?

- If I poured 1 pound of water into a tall, narrow vessel that occupies 1 square inch of floor space

- I would fill that vessel to 2.31 feet of elevation

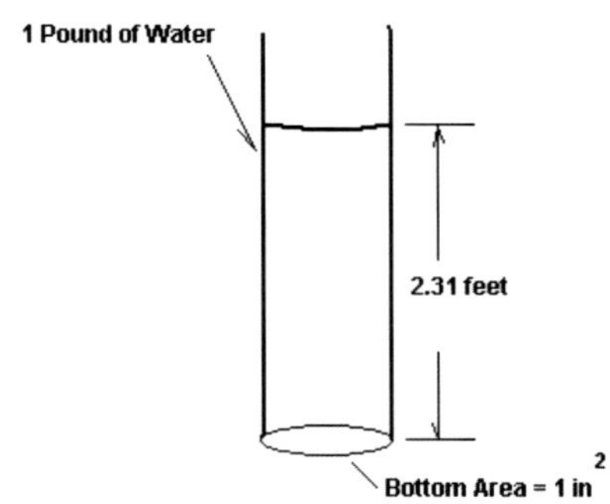

Other Examples

- Most communities will have an elevated tank of ambient water that supplies water pressure to the communities and neighborhoods below the tank. If the water in the tank is 150 feet above a kitchen faucet in one of the homes, what is the water pressure at the faucet (assuming no other influences on pressure)?

- Answer: 150 ÷ 2.31 = 65.8 psi

Other Examples

- Most communities will have an elevated tank of ambient water that supplies water pressure to the communities and neighborhoods below the tank. If the water in the tank is 150 feet above a kitchen faucet in one of the homes, what is the water pressure at the faucet (assuming no other influences on pressure)?
- Answer: 150 ÷ 2.31 = 65.8 psi .
- There would be 66 psi of water pressure available at the kitchen faucet, until someone opens the faucet and water flows. As the faucet is opened and water begins moving through the pipes, there would be a slight pressure drop due to friction between the water and the pipe's internal walls

Head or Pressure

- Pumps develop differential head, or differential pressure .
- This means the pump takes suction pressure, adds more pressure (the design pressure), and generates discharge pressure. So, the discharge pressure is equal to the suction pressure plus the pump's design pressure

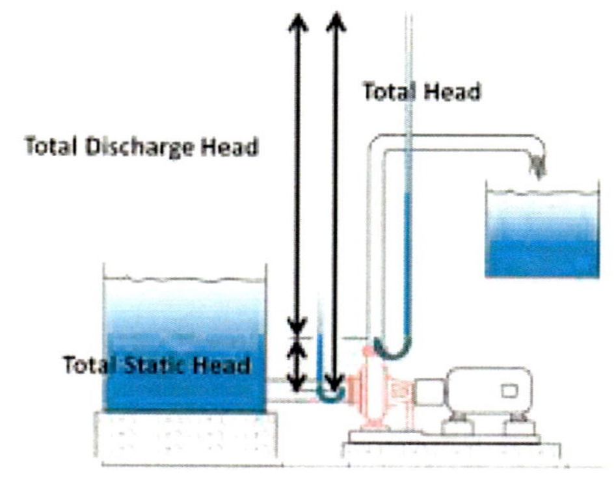

Figure 3: Total Head of a Centrifugal Pump

Head or Pressure

- The discharge pressure of the pump should be approximately equivalent to the total dynamic head (TDH) required by the system (tanks, pipes, elbows, valves, flanges and fittings).
- To monitor and control your pump, your pump should have a suction pressure gauge and a discharge pressure gauge installed on the pump
-

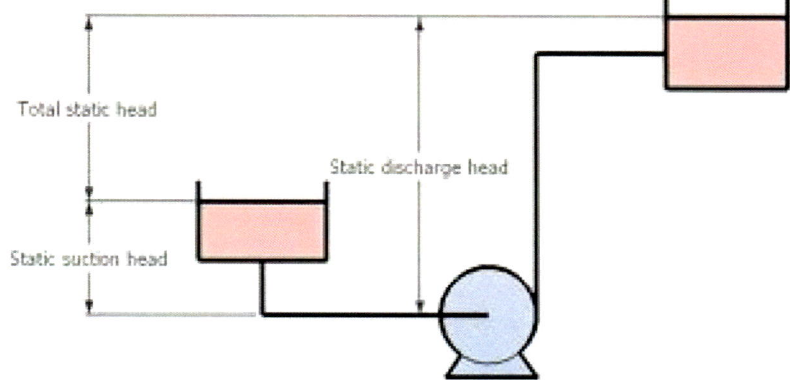

Discharge Pressure

- Let's say your pump is designed to develop 40 psi
- Let's say there are 3 psi of pressure in the liquid as it arrives into the pump
- The suction pressure gauge will read 3 psi
- The pump is designed to add 40 psi of pressure. The discharge gauge would read 43 psi. The differential is 40 psi

Pump Gauges

- Operating a pump without gauges is like driving a car without a dashboard control panel
- If the pressure entering the pump is 25 psi, the discharge gauge will read 65 psi. The differential is 40 psi
- After we control and monitor the differential pressure across the pump, the pump calms down and behaves. The discharge gauge is useless without the suction gauge. Remember, it is the differential pressure, or differential head

Pump Performance Curve

- The pump curve describes the relation between flowrate (GPM) and head (Feet) for the actual pump
- Increasing the impeller diameter or speed increases the head and flow rate capacity - and the pump curve moves upwards

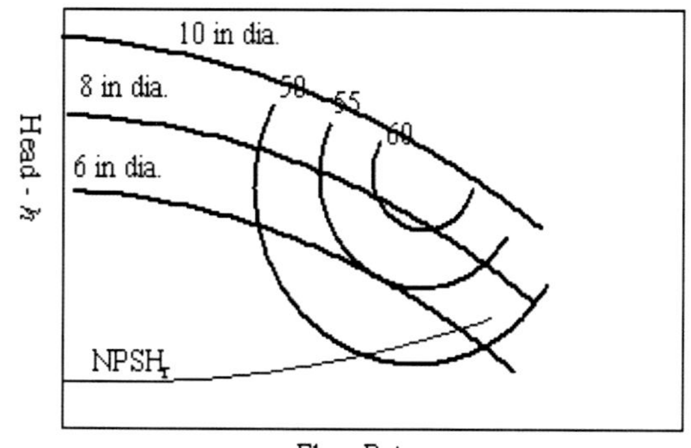

Series Connection

When two (or more) pumps are arranged in serial their resulting pump performance curve is obtained by **adding their** heads at the same flow rate as indicated in the figure below

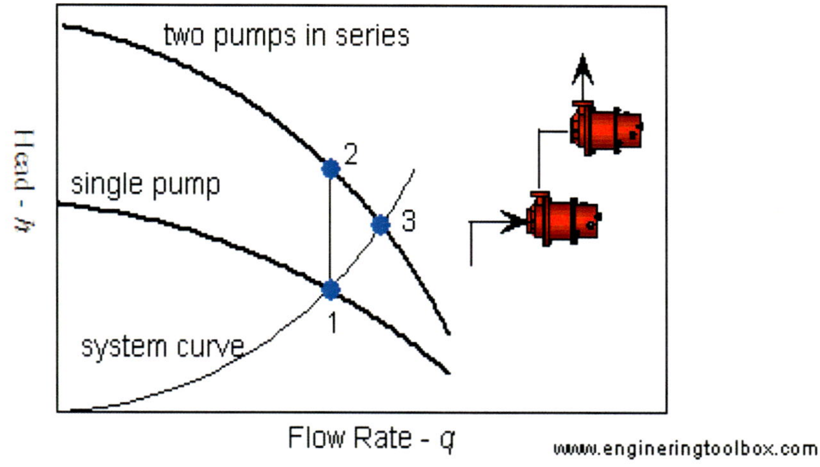

Series or parallel

- The head capacity (or Voltage) can be increased by connecting two or more pumps in series

- The flow rate capacity (Intensity) can be increased by connecting two or more pumps in parallel

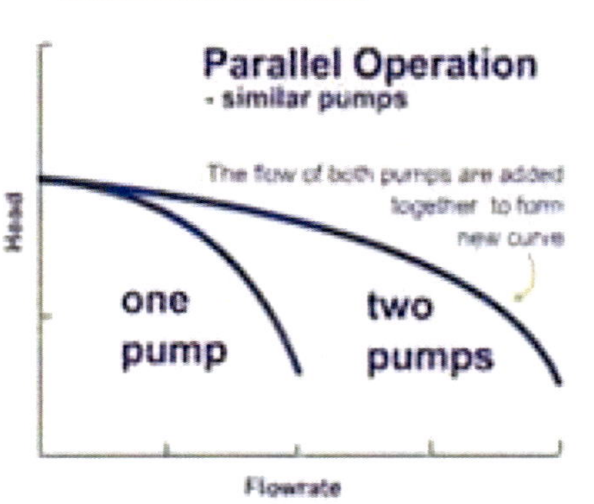

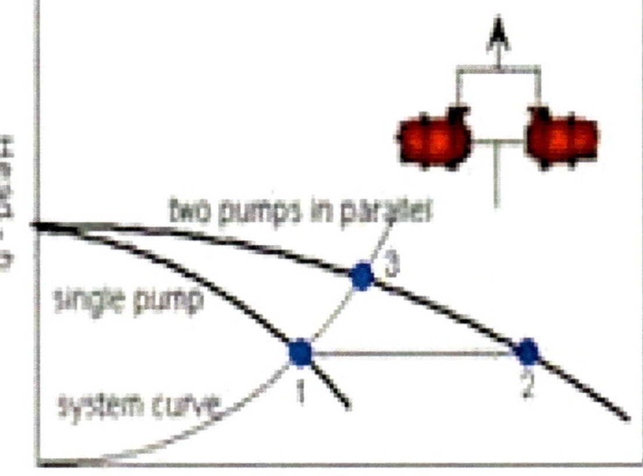

Series Connection

- Centrifugal pumps in series are used to overcome larger system head loss than one pump can handle alone
- For two identical pumps in series the head will be twice the head of a single pump at the same flow rate - as indicated in **point 2**

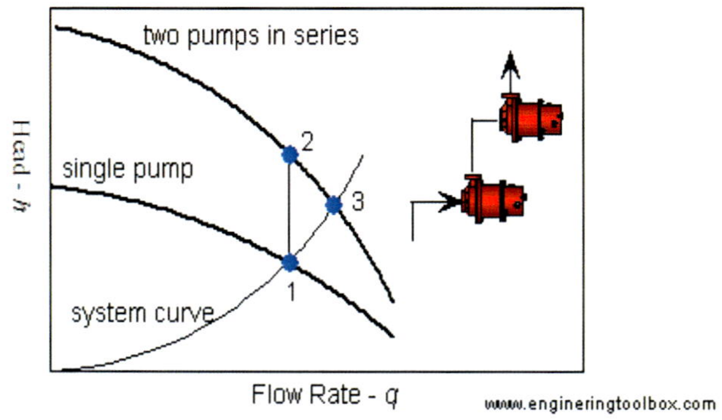

Series Connection

In practice the combined head and flow rate moves along the system curve to point 3.

- **point 3** is where the system operates with **both pumps** running
- **point 1** is where the system operates with **one pump** running

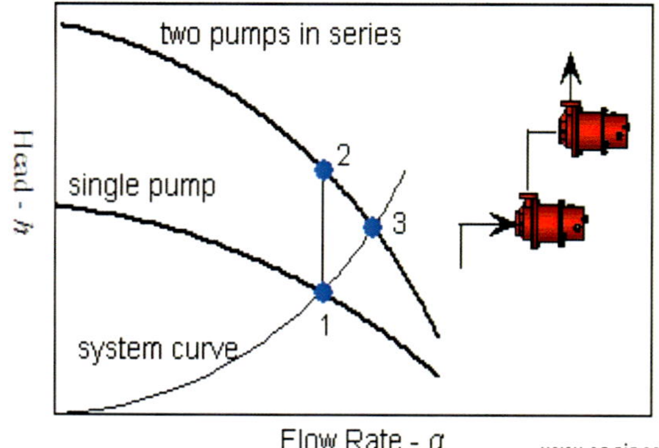

Parallel Connection

- When two or more pumps are arranged in parallel their resulting <u>performance curve</u> is obtained by **adding their flowrates** at the same <u>head</u> as indicated in the figure below .

- Centrifugal pumps in parallel are used to overcome larger volume flows than one pump can handle alone.

  - for two identical pumps in parallel, and the head is kept constant, the flowrate doubles as indicated with **point 2** compared to a single pump

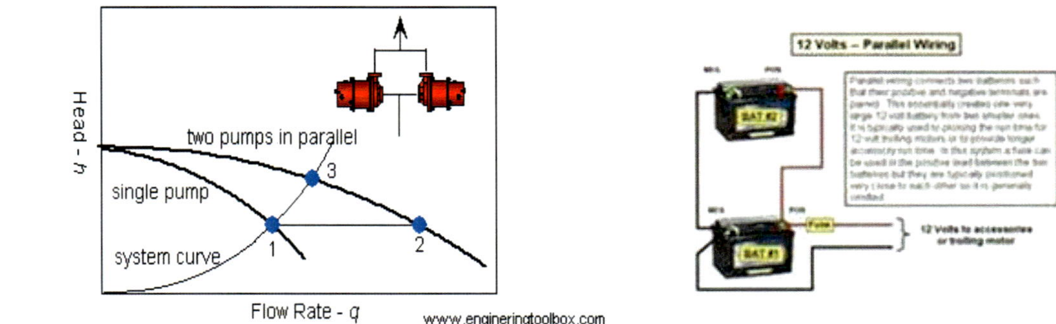

Parallel Connection

- In practice the combined head and volume flow moves along the system curve as indicated from 1 to 3.

- **point 3** is where the system operates with **both pumps** running

- **point 1** is where the system operates with **one pump** running

- In practice, if one of the pumps in parallel or series stops, the operation point moves along the system resistance curve from point 3 to point 1 - the head and flow rate are decreased.

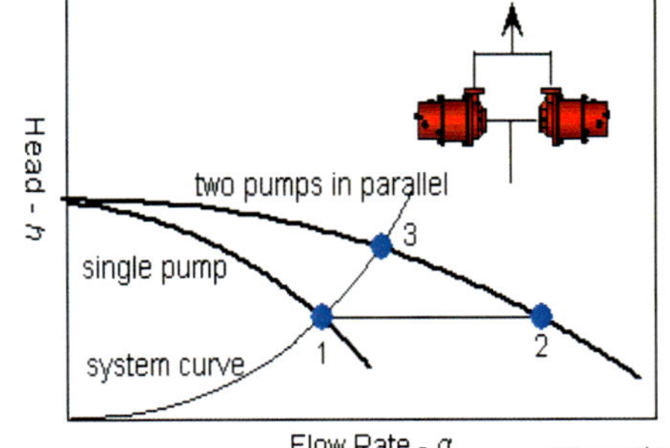

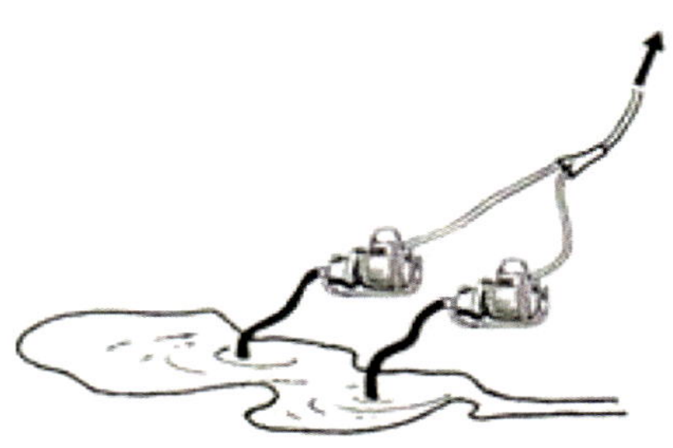

Series Pumps connections

- Centrifugal pumps are connected in series if the discharge of one pump is connected to the suction side of a second pump
- Some things to consider when you connect pumps in series:
  - Both pumps must have the same width impeller or the difference in capacities (GPM or Cubic meters/hour.) could cause a cavitation problem if the first pump cannot supply enough liquid to the second pump.
  - Both pumps must run at the same speed (same reason).
  - Be sure the casing of the second pump is strong enough to resist the higher pressure. Higher strength material, ribbing, or extra bolting may be required.
  - The stuffing box of the second pump will see the discharge pressure of the first pump. You may need a high-pressure mechanical seal.
  - Be sure both pumps are filled with liquid during start-up and operation.
  - Start the second pump after the first pump is running

Parallel Pumps Connections

- Pumps are operated in parallel when two or more pumps are connected to a common discharge line, and share the same suction conditions .
- Some things to consider when pumps are operated in parallel:
  - Both pumps must produce the same head this usually means they must be running at the same speed, with the same diameter impeller .
  - Two pumps in parallel will deliver less than twice the flow rate of a single pump in the system because of the increased friction in the piping .
  - If you run a single pump only, it will operate at a higher flow rate (A) than if it were working in parallel with another pump (B)
-

Pipes and Fittings

- Pipes play principal roles in your system, carrying fresh water to your fixtures and leading wastewater away from your home.
- Pipe fittings allow these pipes to function effectively by linking, turning, and branching them out, and even altering their diameters in order to meet your boat's specific requirements

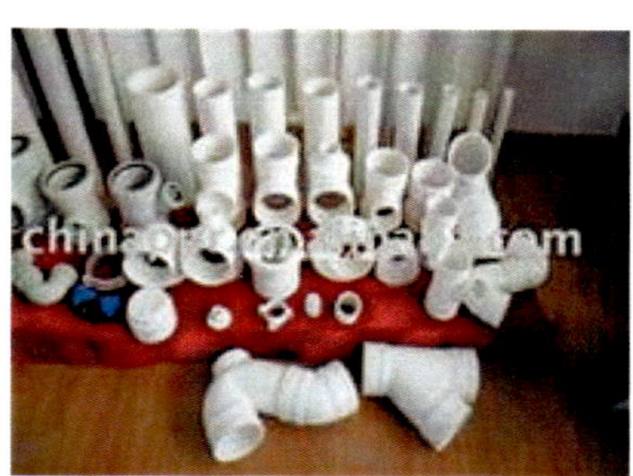

steel pipes   copper pipes   plastic pipes

Materials

- Materials shall be corrosion resistant , non toxic and shall not render the water non-potable, or impart any undesirable taste, odor, or color
- Acceptable materials include:
  - Copper alloys
  - Stainless steel 300 Series
  - Glass lined Metal
  - Plastic and Rubber that comply with applicable National Sanitation Foundation.

Pipe Materials

- Chlorinated Poly (Vinyl Chloride) (CPVC) is a thermoplastic pipe and fitting material made with CPVC compounds meeting the requirements of ASTM and ABYC.
- CPVC piping **which is suitable for hot and cold water distribution** has a 400 psi pressure rating at room temperature, and a 100 psi pressure rating at 180 F

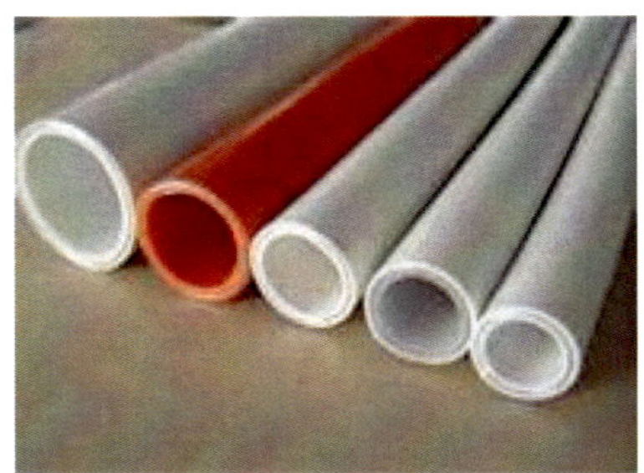

Pipes and Fittings

- The potable water system shall be designed and installed so that the potable water is totally separated from any contact with water used for other purposes .
- Light Blue is the color recommended for potable water system identification if color identification is desired for piping and components

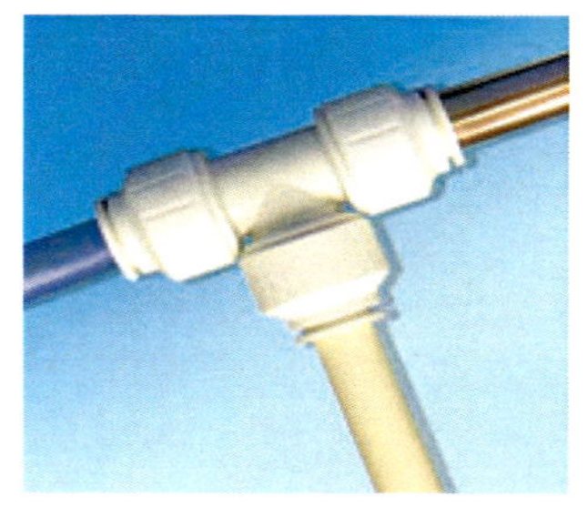

Potable Water System

- The potable water system shall be designed and installed so that the potable water is totally separated from any contact with water used for other purposes .

- Light Blue is the color recommended for potable water system identification if color identification is desired for piping and components

PVC Tubing - Clear – Series/150

- Size Range – 3/16" to 1 1/2"
- This hose is primarily used for non-pressurized potable water system supply and drain lines (sink, shower, icebox)
- This hose **is NOT** for use in hot water or pressurized water systems

Polyester Reinforced PVC Tubing - Series / 168

- Size Range – 1/4" to 1 1/2" 6.4mm – 38.1mm.
- This hose is primarily used for pressurized water system supply (hot or cold), wash down pump connections and head intake
- It has a polyester braid reinforcement and 150 psi maximum burst pressure
-

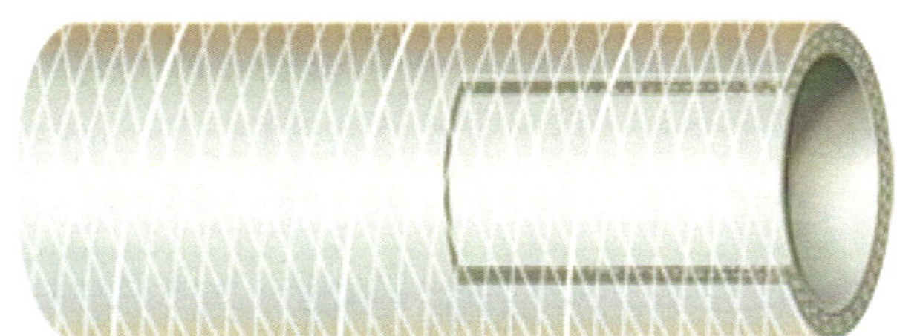

Polyester Reinforced PVC – Red Tracer

- Size Range – 1/4" to 1 1/2" 6.4mm – 38.1mm .
- This hose is primarily used for pressurized water system supply (hot or cold), drain lines (sink, shower, icebox), wash down pump connections and head intake.
- It has a polyester braid reinforcement and 250 psi burst pressure

Polyester Reinforced PVC – Blue Tracer

- Size Range – 1/4" – 1 1/2" 6.4mm – 38.1mm .
- This hose is primarily used for pressurized water system supply (hot or cold), drain lines (sink, shower, icebox), wash down pump connections and head intake.
- It has a polyester braid reinforcement and **more than 250 psi burst pressure**

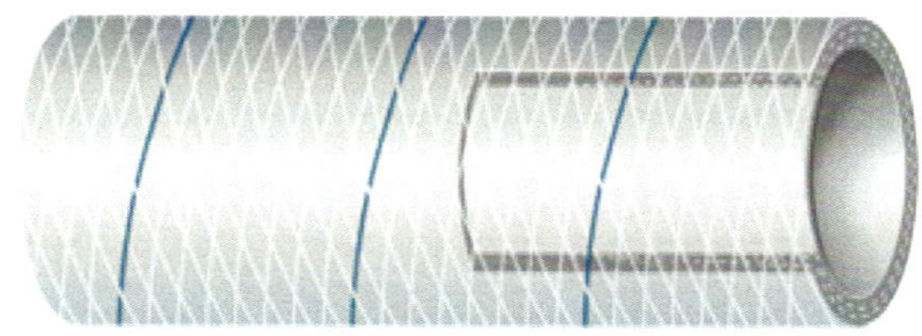

Vac Extra Heavy Duty-FDA – Series/148

- Size Range – 5/8" to 1 1/2" .
- This Hose has a smooth white FDA grade PVC with vinyl helix.
- It is rated for full vacuum and is resistant to mild chemicals, permeation, and saltwater.
- This hose is primarily used for bilge pump intake and discharge, drains (sinks, shower, and scupper), livewell intake and drain, potable water tank fill, and sanitation (toilet to holding tank, intake, self draining lines vent and pump out).
- This hose is resistant to odor permeation. Do not use glycol based antifreeze in hose.

Vac With FDA Liner– Series/145

- Size Range – 1 1/4" to 1 1/2"   .
- Shields Vac with FDA Liner is PVC supported by **vinyl covered steel wire**.
- It has a very flexible FDA formula liner and is resistant to mild chemicals and salt water.
- This hose is primarily used for diaphragm style hand operated bilge pump intake and discharge, drains (sink, shower, scupper), and potable water tank fill.
- Do not use glycol based antifreeze in hose and do not attach to thru hull fitting below water line .

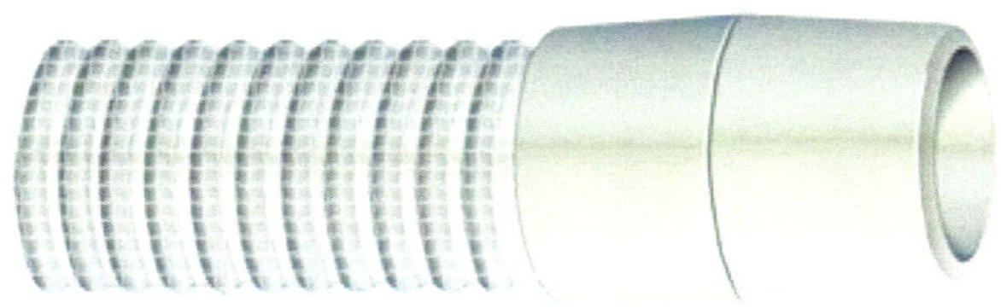

Livewell Hose - Series No. 149

- Size Range – 5/8" to 1 1/2"   .
- Shields Livewell Hose has a black smooth cover with vinyl helix. It is rated for full vacuum and is resistant to mild chemicals and saltwater .
- This hose is primarily used for bilge pump intake and discharge, drains (sinks, shower, and scupper), livewell intake and drain, and sanitation (intake and vent).
- This hose is utilized for below the waterline connections

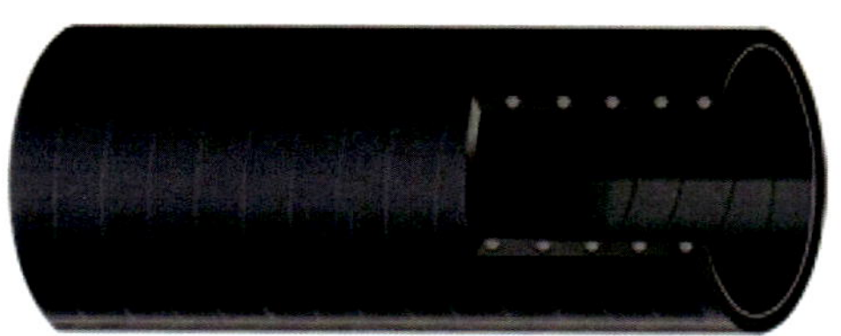

PVC Cutting Tool

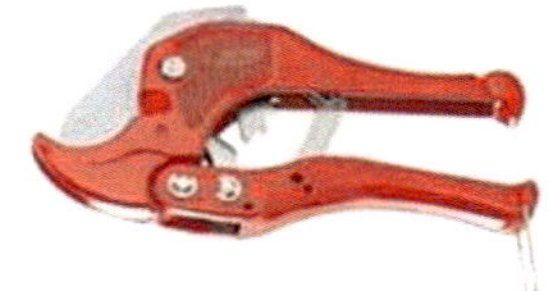

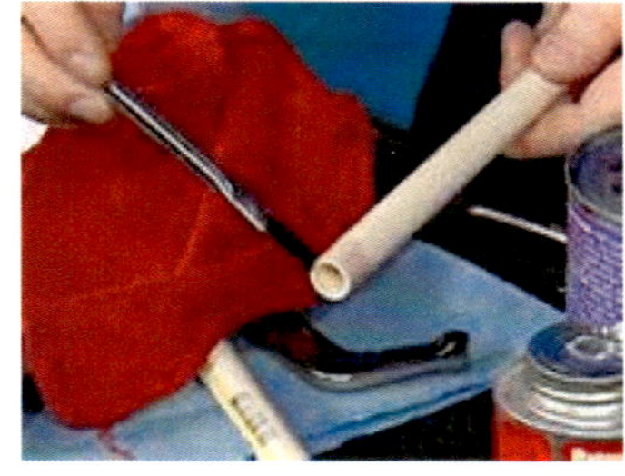

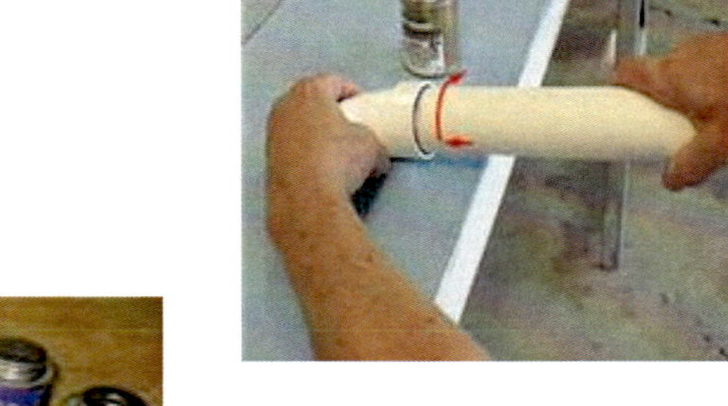

Pressure Switch

- A **pressure switch** opens the electrical circuit to a pump (and turns it off) when the water in the system reaches a pre-set pressure, usually 35-60 psi

- It closes the circuit (and turns the pump back on) when the pressure falls below that point. It is the device that makes on-demand pumps function automatically

Pressure Relief Valve

- A relief valve, set at no more than 150% of the nominal working pressure of the system, shall be provided
- If the potable water system includes a water heater, the pressure relief valve shall be located on the water heater
- If the water heater is equipped with a standard household type relief valve, the water system could be exposed to as much as 400% of the nominal system pressure

Check Valve

- Is a mechanical valve that only permits water flow in one direction. It will also prevent city water pressure (which is higher than the pump can handle) from going into the pump
- If a shore-side connection is provided; a pressure reducing valve and check valve shall be installed at the system inlet connection to reduce the pressure to the nominal working pressure of the boat

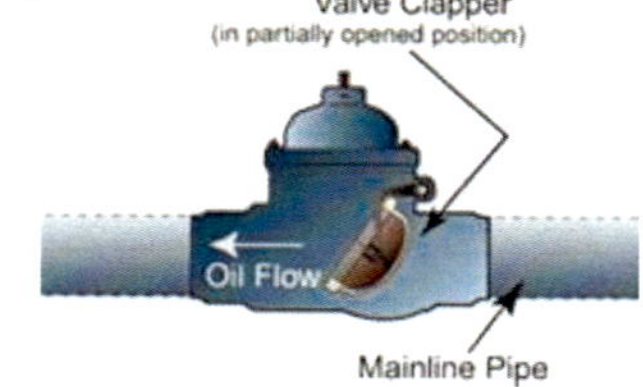

Freshwater Tank

- We think the best tanks are made of thick-walled, high-density polyethylene, but flexible tanks can work in an odd-shaped or inaccessible space

- Tank size depends on the space available and your needs (anywhere from one to ten gallons per person per day might be consumed). Make sure your tank's deck fill has a tight seal and that your tank's vent terminates the boat so your drinking water supply won't be fouled by outside water

The Accumulator Tank

- The Accumulator Tank maintains smooth and steady water flow in your Boat

- Add this compact accumulator tank to your fresh water system just after the water pump

- Pre-charged pressure is 30 psi; maximum operating pressure is 125 psi

Marine Water Heaters

The water heaters can be operated at the dock by AC power or underway by drawing heat from your engine's cooling water. Even a small, 6-gallon heater is enough to provide hot water for washing dishes or taking a short shower

Water Heaters

- Pressure relief valves on hot water heaters need to be checked as well. A typical household variety valve will hold pressure to about 125 PSI
- They may be equipped with a check valve at the water inlet to the tank, located at the bottom of the tank, it will need to be removed as part of the winterization process to ensure that the tank can be drained fully and flushed as needed

Water Heaters

- With respect to the shorepower service hook-up to these water heaters is to be absolutely certain the grounding (green) wire is connected to the case
- Some Heaters are equipped with anodes , these may need servicing periodically
- Also the technician should remember to tag the circuit breaker for the water heater to be certain that no one activates the heating element while it is drained over the winter

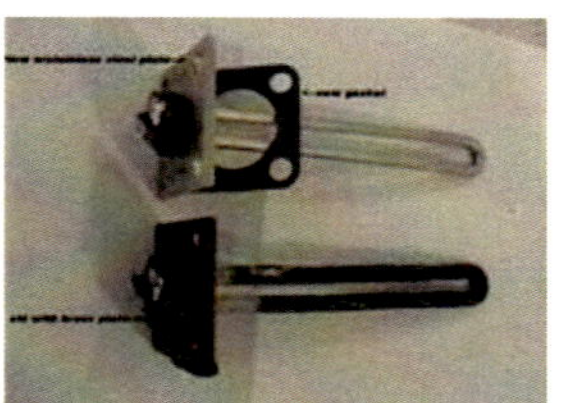

Solar Water Heater

Other modern option is the solar water heater

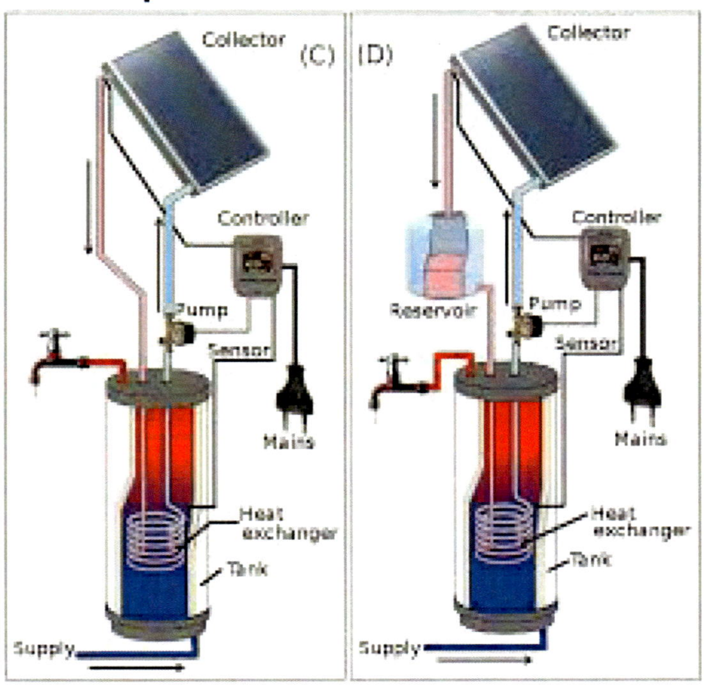

Faucet/Shower

Faucets are the ultimate terminus for water system lines. Mixer faucets require two connections, one from the cold side to the supply line from the pump and the other from the hot side to the water-heater outlet

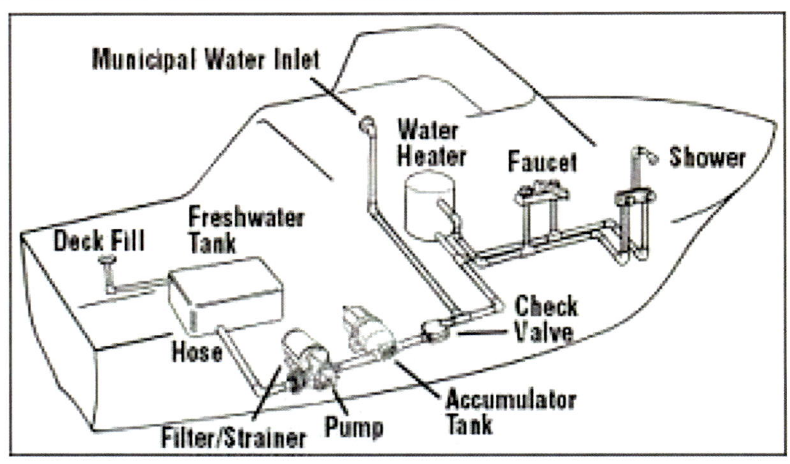

Faucet/Shower

Shower connections are identical to faucet connections. The only difference is that rather than delivering the water through a spigot, the water is delivered through a pipe or hose to the shower head

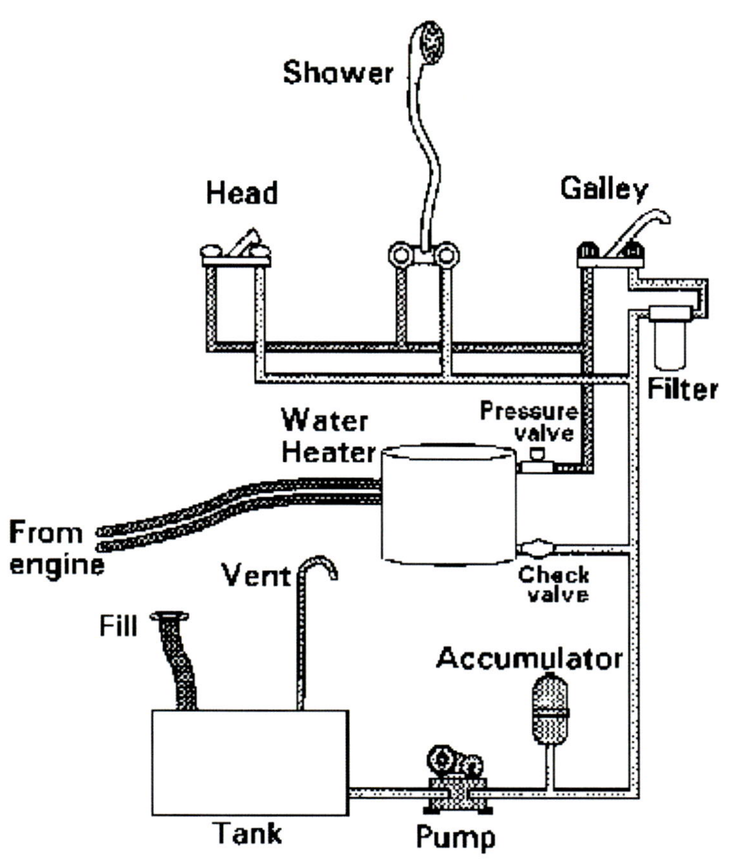

Sump Pumps

Specifications Pump body: Thermoplastic **Shaft seal:** Lip seal **Shaft:** Stainless steel (SS2343) **Voltage:** 12 V DC **Hose size:** L650: 3/4" **Amperage:** 3.2 amp **Fuse:** 5 amp

Sump Pumps

Clean the filter regularly. If using the shower regularly consider changing from bar soap to liquid body wash. Bar soap is "sticky" and tends to gum things up faster - like the float. Body wash is much less gummy

Marine Sanitation Devices

Marine Sanitation Devices (MSD)

- Recreational boats are not required to be equipped with a toilet
- If a toilet is installed, it must be equipped with an operable Marine Sanitation Device (MSD) that is certified by the Coast Guard
- Installed toilets that are not equipped with an MSD, and that discharge raw sewage directly over the side, are illegal
- Portable toilets or "porta-potties" are not considered installed toilets and are not subject to the MSD regulations

Marine Sanitation Systems

- **Type One MSD's** typically discharge treated waste directly overboard
  - This system reduce bacteria and discharge no visible floating solids
- **Type two MSD** system is similar to type one but with a better job of treating waste
  - These are typically installed in boats over 65', but can be used on any vessel outside of a non discharge zone
- **Type Three MSD** system uses a head connected directly to a holding tank
  - Some systems use "Y –Valves" that allow waste flow to be diverted directly overboard (outside 3 mile discharge limit)
  - Holding tanks can be emptied at many marinas or state run pump-out stations

The Marine Environment

- Portable Toilets are subject to disposal regulations that prohibit the disposal of raw sewage within territorial waters (3 mile limit), the Great Lakes, or navigable rivers

- This symbol is used to show boaters where onshore pump out services are located. You will need to locate a marina displaying this symbol if you use a portable toilet or Type III MSD holding tank on your boat

No Discharge Zones

- A boat can be equipped with any type of MSD permitted under the regulations .

- However, whenever a boat equipped with a Type I or Type II MSD (these types discharge treated sewage) is operating in an area of water that has been declared a No Discharge Zone, the MSD cannot be used and must be secured to prevent discharge.

- No Discharge Zones are areas of water that require greater environmental protection and where even the discharge of treated sewage could be harmful.

Type I Treatment System

- A treatment system (Type I MSD), offers nearly the same plumbing simplicity as direct discharge

- Connect the head's discharge hose to the inlet side of the treatment unit, and connect the outlet side to the discharge through-hull

- Some onboard treatment systems do a better job than municipal sewage plants

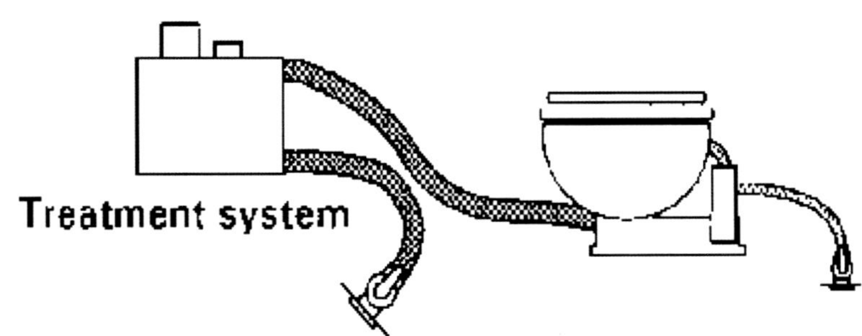

MSD Type II

- This type of sanitation system is divided in three treatment stages; aeration, clarification, and disinfection .
- In the aeration chamber (stage 1), the bacteria grow and multiply using the sewage as their food supply. This action reduces the quantity and size of the solid matter

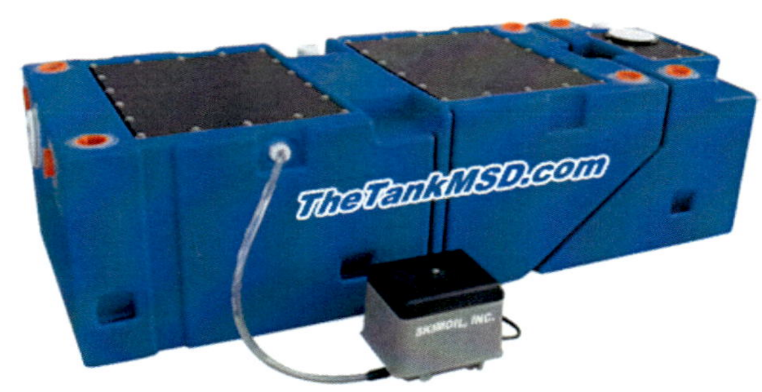

MSD Type II

- In the clarification chamber (stage 2), the bacterial floc is separated from the treated solid matter. The treated water is clear and free from solids.
- The liquid must be disinfected prior to discharge overboard to kill any disease-causing bacteria
- Disinfection is accomplished in the clarification chamber (stage 3)

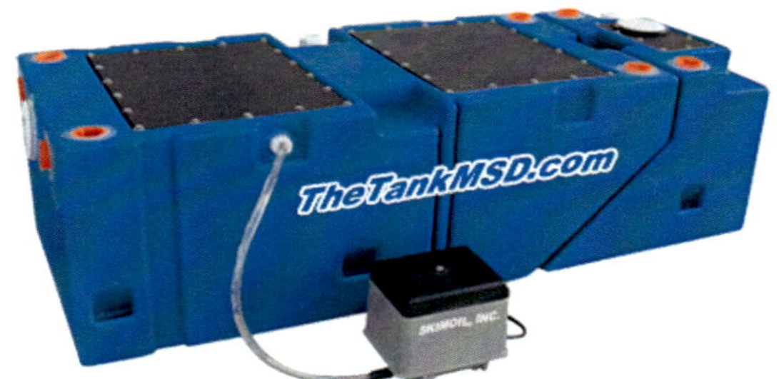

MSD Type II

- Flow through these three stages is caused by direct displacement.

- When new sewage flows into the aeration chamber, an equal volume flows through the clarification chamber. This volume, in turn, displaces an equal volume from the clarification chamber into the disinfection chamber, and overboard.

- No internal sewage pumps are necessary.

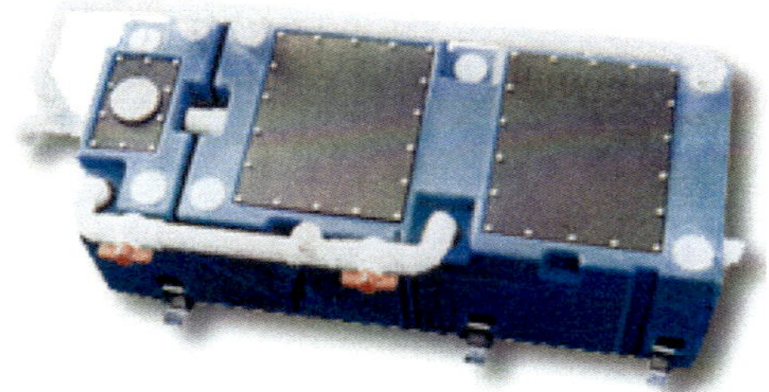

DISCHARGE OF RAW SEWAGE

The boundaries of U.S. territorial waters are marked on some nautical charts. Changes to the boundaries are published in Coast Guard Local Notices to Mariners

Discharge of Raw Sewage

It is illegal to discharge raw sewage from a boat in territorial waters (within the 3 mile limit), the Great Lakes, and navigable rivers. However, a Y-valve may be installed on any MSD to provide for the direct discharge of raw sewage when the boat is outside U.S. territorial waters

Discharge of Raw Sewage

The valve must be secured in a closed position while operating in U.S. waters

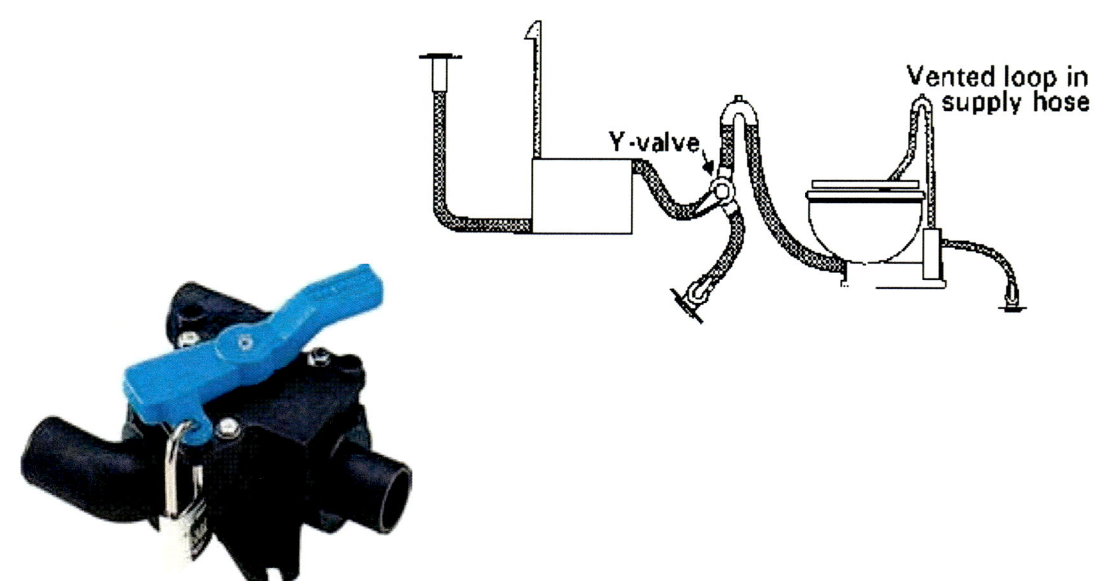

Coastal Use

- If your boat never leaves inland or coastal waters, connecting the head directly to the tank is your only legal option
- The tank itself will require two additional connections, one to a pump-out fitting on deck and the other to an outside vent fitting to prevent a build-up of explosive gas inside the tank
- All three of these hoses must be sanitation hose to resist gas permeation

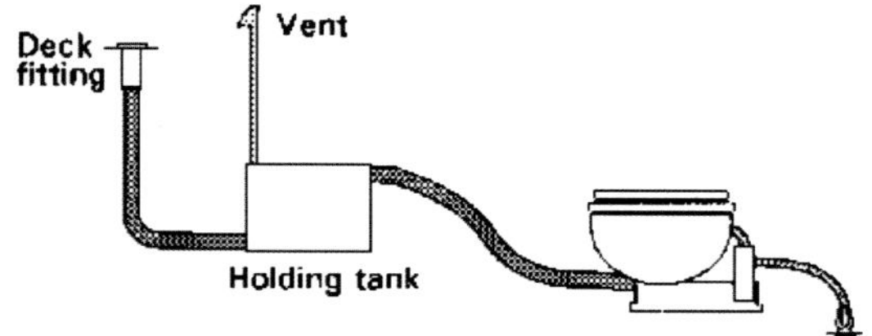

Waste Discharge seacock

Larger Yachts and cruisers will probably have a more sophisticated power supply on board and should be able to cope with the additional drain of an electric toilet

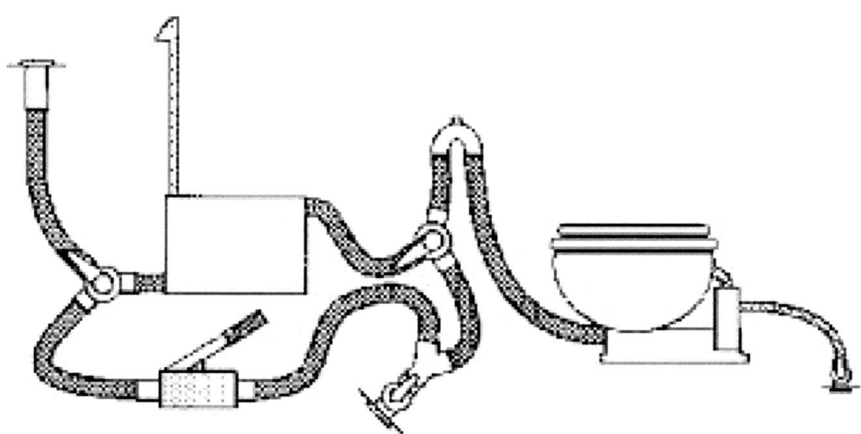

Waste Discharge seacock

Sea going vessels also have the opportunity to discharge overboard when out at sea subject, of course, to restrictions operating in that particular area and may require the provision of a holding tank as well as a waste discharge seacock

Marine Toilets

- The type of system you choose will depend on the type of boat you have, power supply, fresh water capacity and cruising area

- Small yachts or cruisers may not have a sufficient power supply to cope with the demands of an electric toilet and therefore a manual sea toilet will be required

Toilet System

- Water supply will depend on fresh water and holding tank capacities

- Small yachts and cruisers may only have small water and holding tanks therefore a raw water supply with the facility to pump to a holding tank or overboard will be necessary

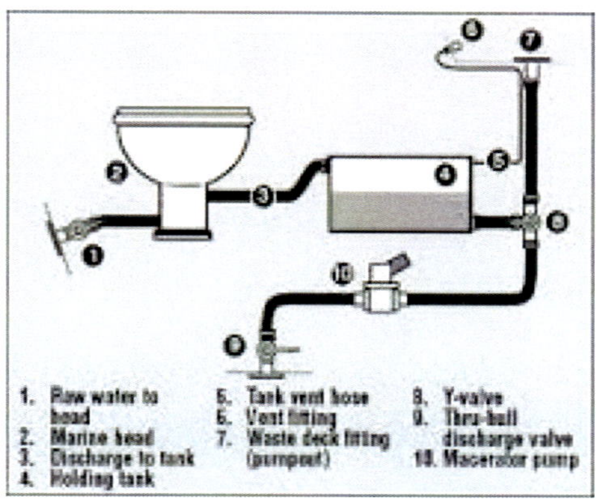

Toilet System

- Larger vessels normally have a substantial fresh water and waste holding tank and can cope with the demands of a toilet system

- The drawback with using a raw water flushing system can be that contaminated water is drawn into the inlet pipework, the toilet bowl and if allowed to stand for a protracted period, will not only discolor the bowl but also lead to noxious smells which can deter even the most hardy of mariners

- **Fresh water flushing will prevent odors from occurring and keep the crew happy**.

Manual Toilet

Manual toilets are normally considered for use where space, power and weight are important factors and where water consumption is not of prime importance

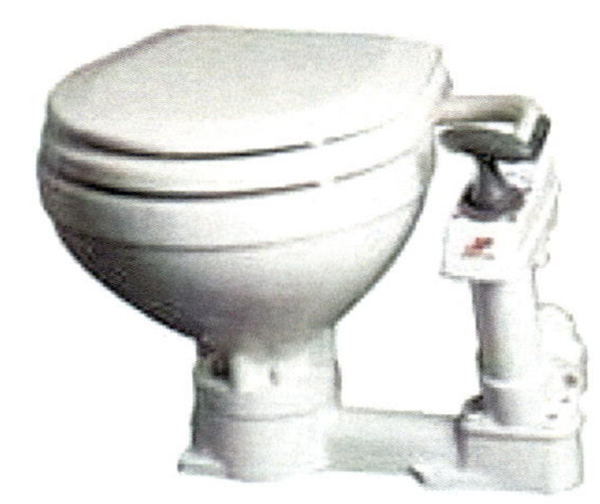

Gravity Discharge

Gravity discharge toilets offer the simplest solution to pump-out systems with the toilet sitting directly over the waste tank giving very little opportunity for problems although smell will become an issue if the tank is not pumped out on a regular basis

Macerator and Centrifugal

- Macerator and centrifugal or compact toilets are the closest in design to a normal household toilet

- They can be freestanding and will pump a considerable distance to a remote waste tank. Many of these toilets have eco flush facilities therefore reducing the amount of water used, a boon where water and waste capacity is limited

Vacuum Toilets

When the vacuum has stabilised, usually between 30 to 45seconds the discharge mechanism can be operated and the vacuum removes the contents of the bowl at high speed, pulls it through the vac tank and up to the vac pump from where it is then pushed into the waste tank

Macerator Pump

The Jabsco macerator is the ideal answer to simple holding tank pump-outs. Flow rate of 45 litres per minute empties a 120 litre holding tank in only 3 minutes. Heavy duty motor discharges up to 1,000 litres in one operation

Macerator Pumps

- Electric macerator pumps can pump holding tanks and certain types of toilets and some can be used as macerators

- Macerator Pumps are very compact and lend themselves to installations short on space

Macerator Pump

- Macerator pumps break up solids, which minimizes clogging. They can be used to pump into a holding tank or to empty it overboard depending on the macerator pump placement
- Less water is required for flushing if a macerator pump is installed on the input side of the holding tank .
- Macerator pumps can require up to 10 or 15 amps of power and cannot be run dry .

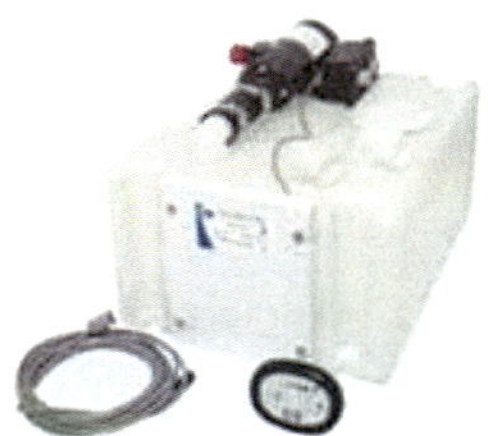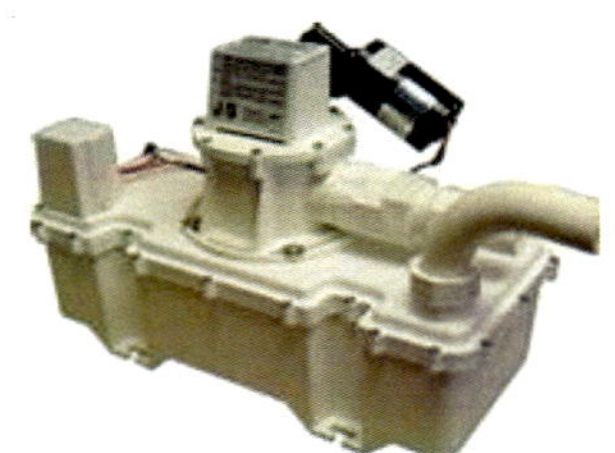

Macerator Pump

- The function of the pump is to suction the solids and liquids from the lines connected to the holding tanks and grind the effluent with the rotating cutter head down to a small particle size for simple discharge of the waste
- The term macerator refers to the action of the pump that grinds the waste as it enters the inlet of the pump

Macerator Pump

- Installation of a macerator pump is very straightforward. Since the pump is powered by electricity, either DC or AC, the power should be shut off prior to connecting to the supply
- Circuit protection, either an inline fuse or a circuit breaker should be used to protect the circuit

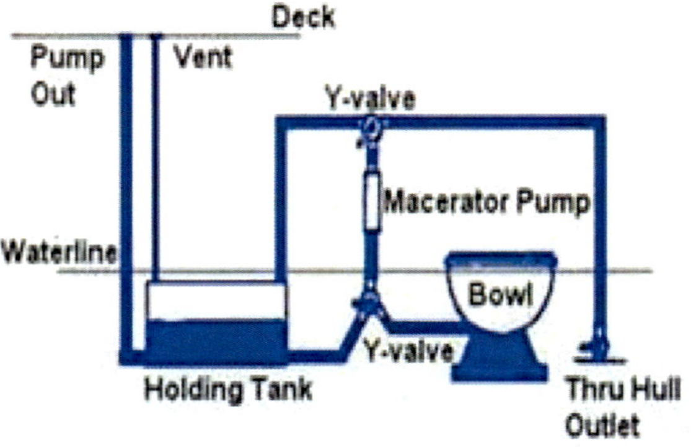

Manual Pump Design

Manual Pump with Charcoal Vent, Deck Pumpout and Ball Valve

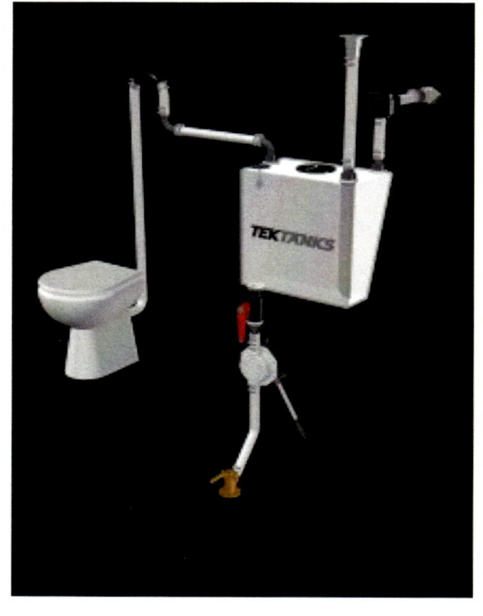

Vented Loop

If your head is mounted below the waterline, or if it moves below when the boat heels, you must have a vented loop in a discharge line that connects to a through-hull fitting

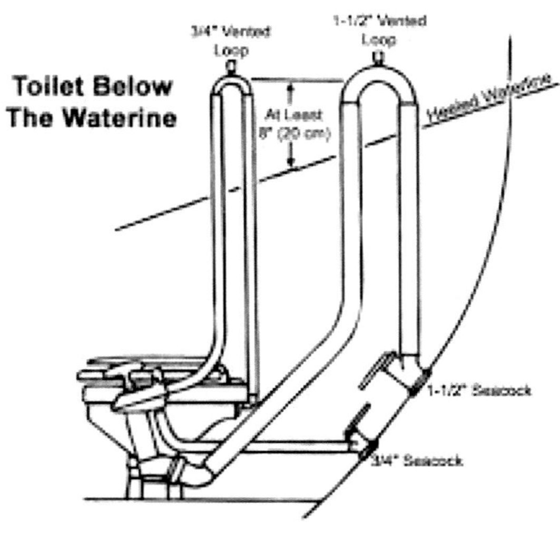

Vented Loop

Mount the vented loop so it will remain above the waterline at all heel angles. Clean the anti-siphon valve regularly to keep it functioning

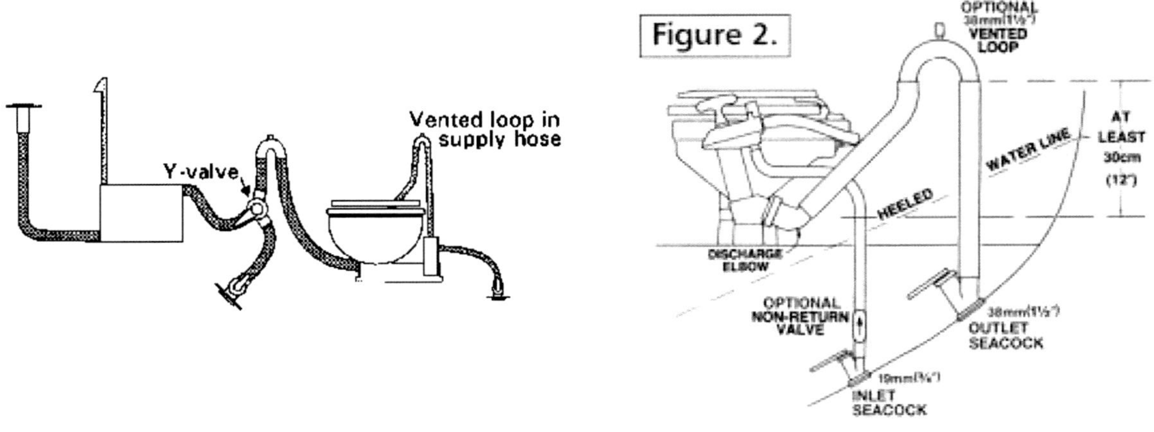

Vented Loop

- A vented loop in the inlet line can interfere with the proper functioning of the head, and its omission poses less risk because of the positive-action valve on the inlet side of the head

- But if you leave the loop out, you must keep the inlet valve in good working order. A screen filter to exclude grass and other debris is highly recommended

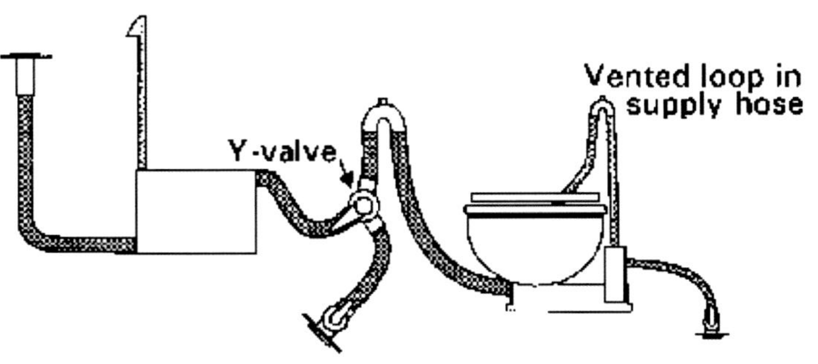

Marine Holding Tank

A marine holding tank receives an excrement and water charge from a toilet and converts such charge into vapor and ash components for efficient and hygienic storage and/or removal

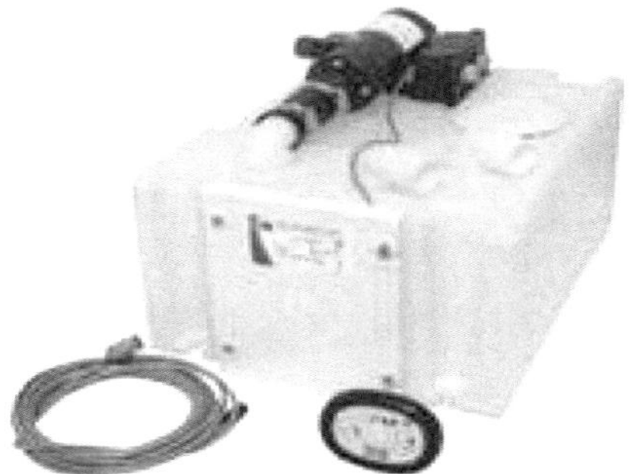

Marine Holding Tank

A holding tank has the advantage of being the only sewage handling method universally acceptable to all governing authorities, and it adds the least complication to sewage handling

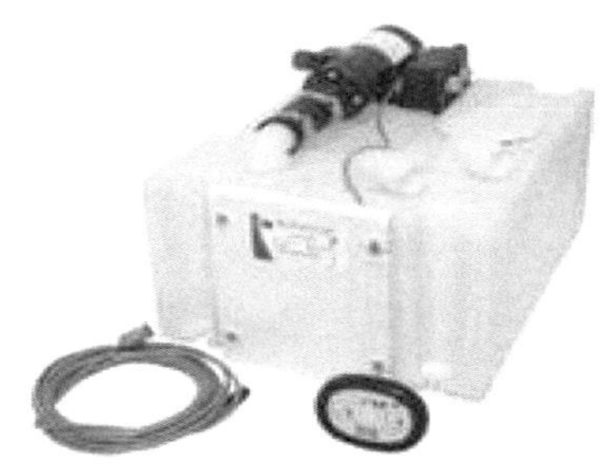

Coastal Use

- If your boat never leaves inland or coastal waters, connecting the head directly to the tank is your only legal option. The tank itself will require two additional connections, one to a pump-out fitting on deck and the other to an outside vent fitting to prevent a build-up of explosive methane inside the tank

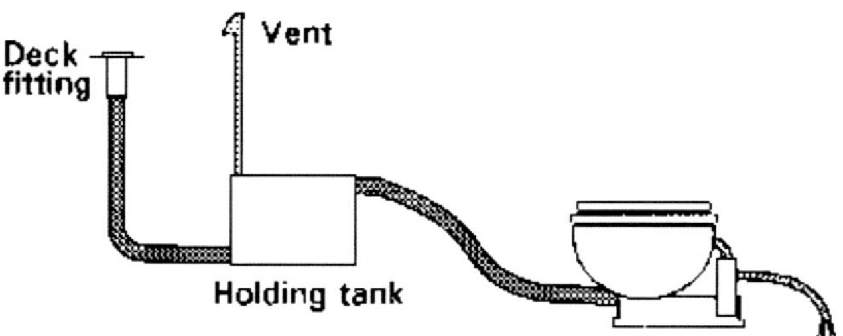

Holding tank Leaks

- First, inspect the holding tank to confirm it's not leaking. This is especially important if the tank is metal and more than a few years old: Acids in urine will eventually eat through metal

- Fill the tank about a quarter of the way full with a hose through the deck pump-out; the pump-out pipe reaches to the bottom of the tank, so the inflowing water will stir up the gunk down there

Holding tank Leaks

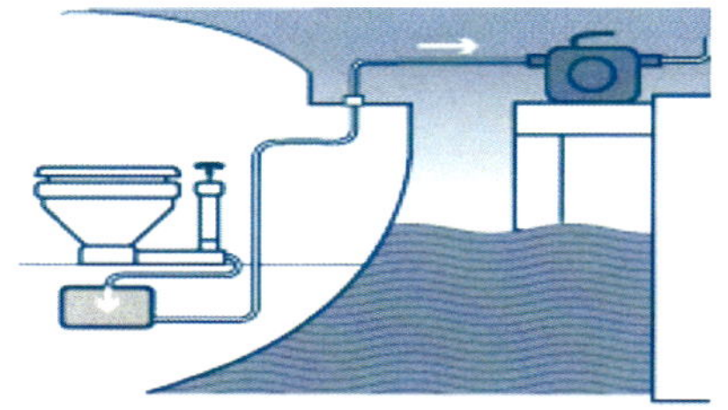

- Now pump the tank, refill it, and pump it again; repeat this process until the discharge is clean. It might take several cycles

- Then treat the tank with a cleaner/deodorizer to kill any remaining whiff

- If you've been living right, cleaning the holding tank might solve your odor problem. If not, the next step is to tackle the hoses, Y-valve, macerator pump, and other potential leak sources. We'll cover this in a future "Around the Yard" column

Holding Tank Monitor

New design electronics provide high reliability and long life with zero current draw except for the short periods required for level checking.
Senders are simple, solidly constructed units with no moving parts, thus no floats to stick.
Hookup consists of simply connecting 12 VDC power and the two-wire senders

Typical Installation

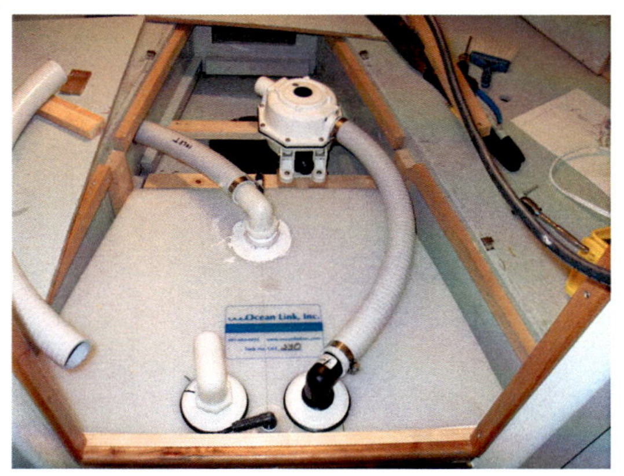

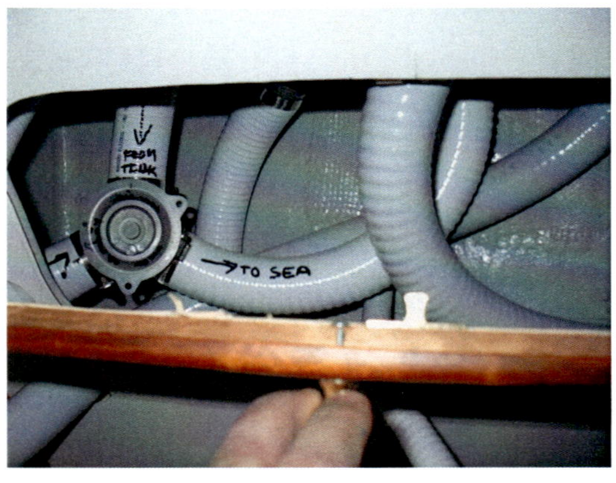

Chapter 9
Fire Protection Systems

- Portable Extinguishers
- Types of Fires
- Fire Extinguisher Ratings
- Marine Fire Suppression System
- Sources of Fires
- Wiring a Fire Suppression System

Fire Protection Systems

- 90 percent of boat fires are likely to begin in the engine room however the other 10 percent comes from appliances with a cycle on and off or from cigarette smoking, and electric space heaters

- While the National Fire Protection Association (NFPA) recommends smoke detectors on boats 26-feet or larger with sleeping areas, ABYC does not require smoke detectors at this time (but there are efforts being made to get this requirement into the standard).

Portable Fire Extinguisher Parts

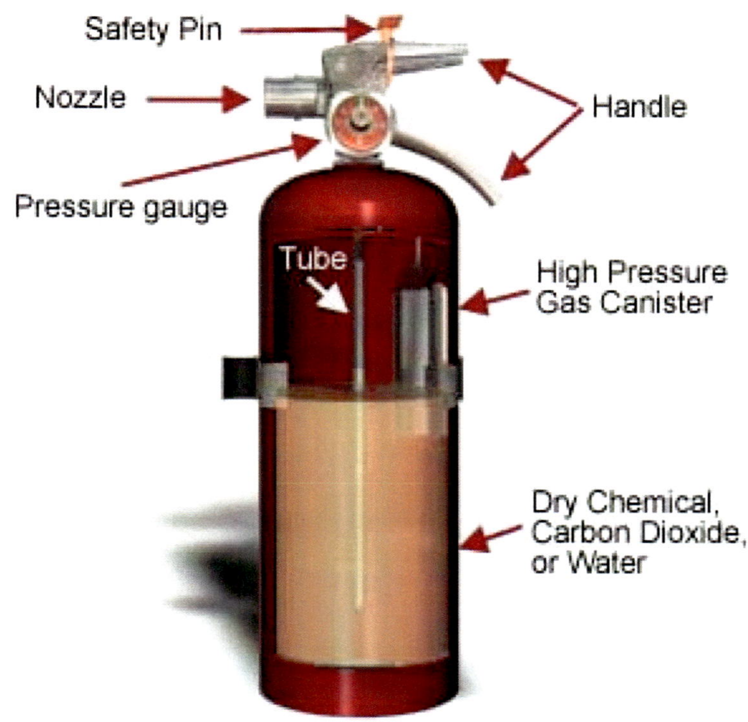

Portable Fire Extinguisher

Portable fire extinguishers apply an extinguishing agent that will either cool burning fuel, displace or remove oxygen, or stop the chemical reaction so a fire cannot continue to burn

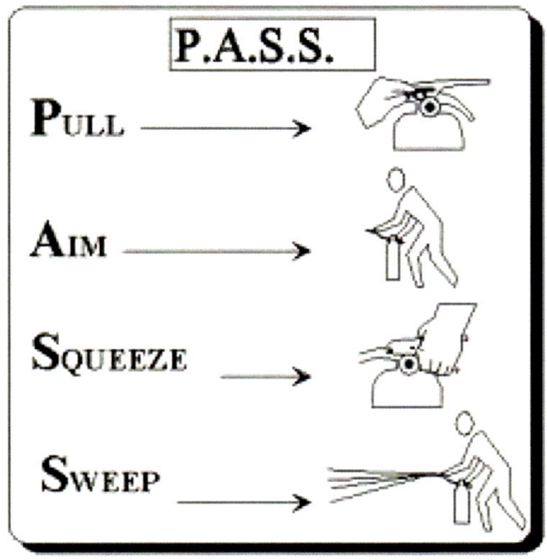

Portable Fire Extinguisher

- Portable fire extinguishers are classified to indicate their ability to handle specific classes and sizes of fires
- Labels on extinguishers indicate the class and relative size of fire that they can be expected to handle
    - Portable extinguishers should be mounted where they are readily accessible.
    - Check the gauge to make sure the extinguisher is still charged.
    - Check the seals to make sure they have not been tampered with.
    - Replace cracked or broken hoses and keep nozzles free from obstruction.
    - Weigh extinguishers to assure that they meet the minimum weight stated on the label

Fire Extinguisher Ratings

Fire extinguishers are classified by fire type. The A, B, C, D rating system defines the kinds of burning materials each fire extinguisher is designed to fight

A		Common Combustibles	Wood, paper, cloth etc.
B		Flammable liquids and gases	Gasoline, propane and solvents
C		Live electrical equipment	Computers, fax machines
D		Combustible metals	Magnesium, lithium, titanium
K		Cooking media	Cooking oils and fats

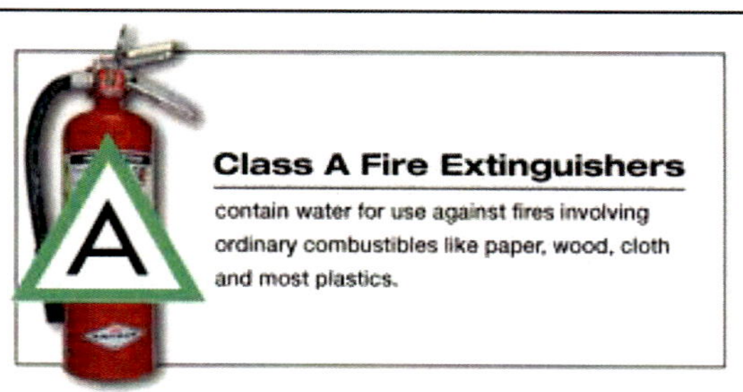

<u>Fire Extinguishers For Class A Fires</u>

- Paper
- Wood
- Textiles
- Plastics

Class A Extinguishers

- Class A extinguishers are used on fires involving ordinary combustibles, such as wood, cloth, textiles and Plastics

- Extinguishers suitable for Class A fires should be identified by a triangle containing the letter "A." If colored, the triangle should be green

Recommendations

- Some boat fires involve burning wood and paper (Class A), but these fires can be put out with water

- Do not use water on gasoline, oil, or electrical fires. Water causes gasoline and oil fires to spread and electrical current is conducted through the water

- Make sure to inspect your fire extinguishers monthly to make sure they are properly stored, charged and undamaged

Fire Extinguishers For Class B Fires

- Oil
- Gasoline
- Kerosene
- Cooking oil

Class B Extinguisher

- Class B extinguishers are used on fires involving liquids, greases, and gases

- Extinguishers suitable for Class B fires should be identified by a square containing the letter "B." If colored, the square shall be colored red

What type of materials would be burning for a fire to be classified as a Class B fire?

- Oils
- Gasoline
- Some Paints
- Grease
- Lacquers
- Solvents
- Flammable liquids

FLAMMABLE LIQUIDS

What type of fire extinguisher is labeled for Class B fires?

- <u>CO2 carbon dioxide fire extinguishers</u> and **dry chemical** fire extinguishers are both labeled for Class B fires.

What does a CO2 fire extinguisher look like?

- This type of fire extinguisher has a hard horn and NO pressure gauge

- They are red and range in size from five to one hundred pounds or larger

What type of fires should never be extinguished with CO2 fire extinguisher?

CO2 fire extinguishers should never be used with fires involving ordinary combustibles (paper, cloth, wood, rubber and plastic) because the fire could continue to smolder and then reignite after the dissipation of the CO2.

<u>Fire Extinguishers For Class C Fires</u>

ELECTRICAL EQUIPMENT

Class C Extinguisher

- Class C extinguishers are used on fires involving energized electrical equipment

- Extinguishers suitable for Class C fires should be identified by a circle containing the letter "C." If colored, the circle should be colored blue

What type of materials would be burning for a fire to be classified as a Class C fire?

- Wiring   **ELECTRICAL EQUIPMENT**
- Fuse boxes
- Energized electrical equipment
- Computers
- Other electrical sources

What type Chemical is used on a fire extinguisher labeled for Class C fires?

Both monoammonium phosphate and sodium bicarbonate are commonly used to fight this type of fire because of their nonconductive properties

Recomendation

For fires involving <u>electrical equipment</u>, Carbon Dioxide (CO2) fire extinguishers and Dry Chemical fire extinguishers are used

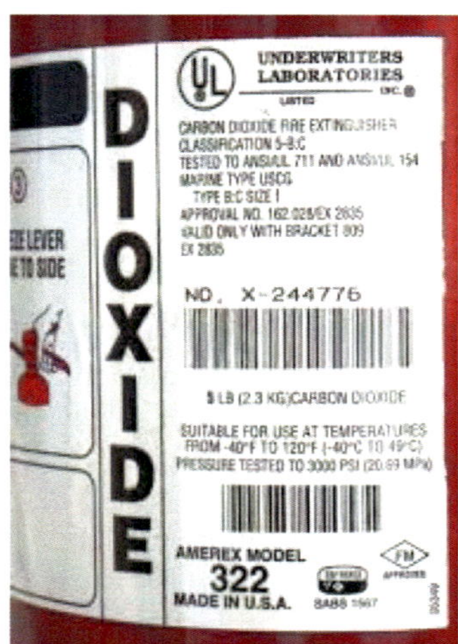

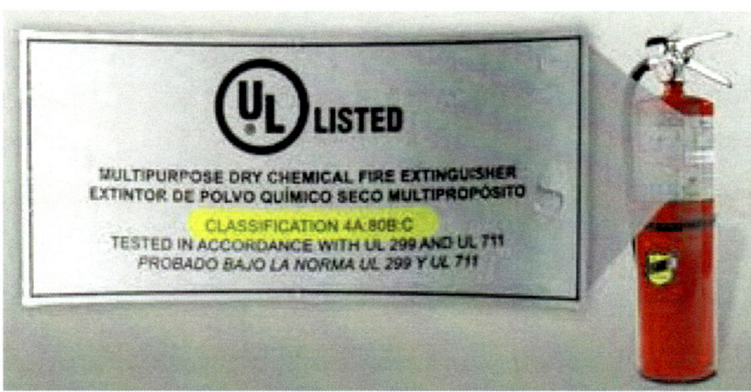

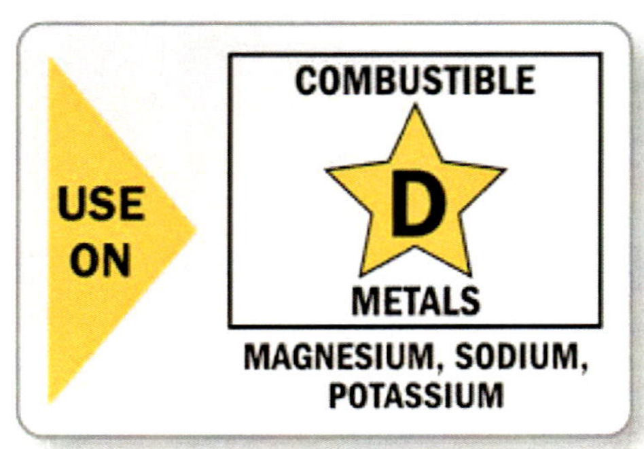

Fire Extinguishers For Class D Fires

Combustible Metals

Class D Extinguisher

- Class D extinguishers are used on fires involving metals such as magnesium, titanium, zirconium, sodium, and potassium

- Extinguishers suitable for fires involving metals should be identified by a five-pointed star containing the letter "D." If colored, the star shall be colored yellow

Which extinguisher is best for which type of fire?

Type of Extinguisher	Class of Fire	Notes:
Carbon Dioxide (CO₂)	B and C	Carbon Dioxide is a class B, C, agent only. Because of the CO2 high pressure, it is not recommended for use on Class A, ember and ash-based fires because of the hazard of spreading the fire when blasting it with the high pressure gas.
Dry Chemical	B and C	The A, B, C dry chemical is not recommended for marine use for two reasons: -It is corrosive -The way in which this agent obtains its class A rating is its ability to melt, seep and encase. This necessitates dismantling of equipment to repair or rebuild.
Foam	A and B	Foam extinguishers are water based and quench Class A fires. They also blanket, smother and separate the vapor layer in Class B fires

Fire Suppression System

The first line of defense when it comes to preventing a fire on board is to make sure that all systems are correctly installed and that there is nothing that can turn a small problem into a large one

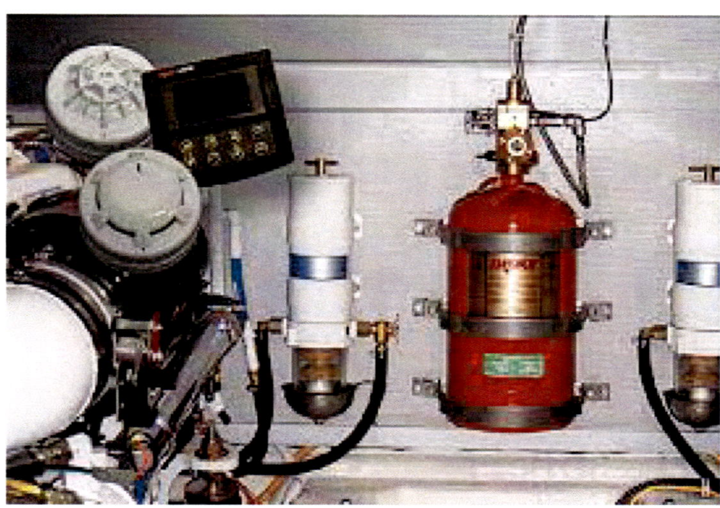

Fire Suppression System

- The ABYC recommends that engine fires be tackled without opening the engine cover or—in larger vessels—the door to the engine room.
- The cover or door maintains a physical barrier between you and the fire and restricts the <u>amount of oxygen</u> that can reach the flames

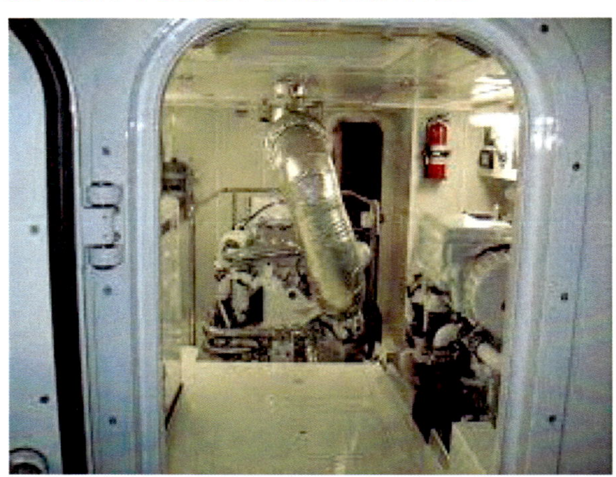

Fire Suppression System

- A pressurized extinguisher bottle containing clean agent FM-200 is placed in the compartment to be protected
- The bottle is sized to protect the volume of the space, and the bottle nozzle includes a fusible link that will melt at 175°F

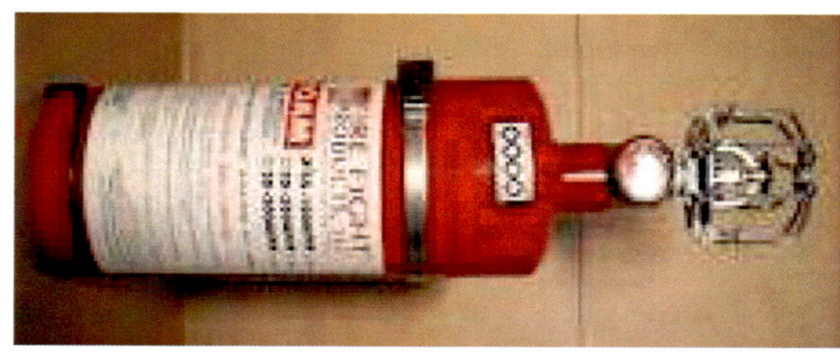

Fire Suppression System

It is a thermal comparison system that senses the engine room temperature, and when it rises above a preset limit, deploys an inert gas to deprive the engine room of oxygen, thus putting out the fire

Fire Suppression System

When 175°F (79°C) is reached, the extinguisher will automatically discharge, releasing the clean agent gas to totally flood the entire space, smothering the fire

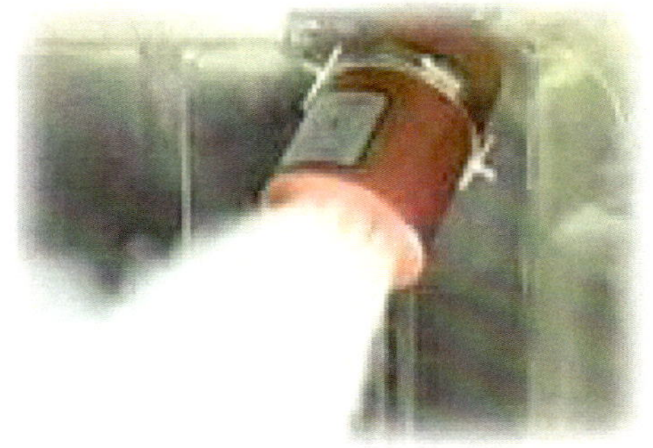

The FM 200

The FM-200 itself is non-toxic and will not leave any residue or damage equipment. However, the FM-200 will displace oxygen and, depending on the fire, can mix with the byproducts of the fire to create a toxic environment – both of which can be dangerous

Fire Suppression System Activation

The systems can be activated either manually or electronically

All automatic systems also include manual actuation pull cables that can be actuated from outside the space

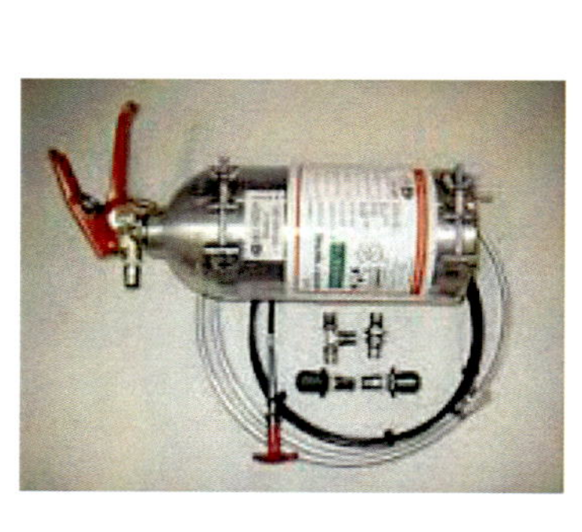

Manual / Automatic   Automatic

Fire Suppression System

- Halon was used in earlier models but environmental concerns have resulted in it being outlawed.
- Modern systems use another gas called heptafluoropropane, although halon is still used
- HFC-227ea is an effective Halon replacement clean agent

Engines Shut Down

- If engines, generators, blowers and some other types of machinery are preset, the automatic system will shut this equipment down when the extinguisher bottle is actuated
- Small control panels for the automatic systems are provided in the pilothouse

The Pilothouse Control Panel

The pilothouse control panel will indicate if the shutdown system is powered, and a dial indicator on the extinguisher bottle itself will show if the bottle is pressurized

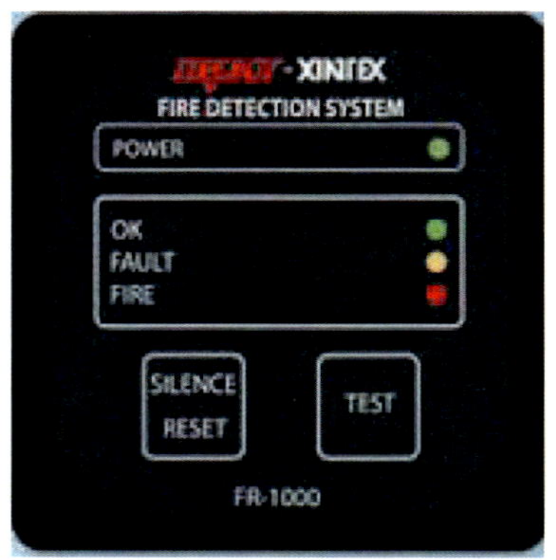

Testing the System

Routine system checks are recommended, and an annual inspection by a professional should be done, which includes removal of the extinguisher bottle so it can be weighed. The shutdown system should be tested yearly to ensure all the relays are working properly

Sources of Fires

- Undersized wires , insufficient circuit protection , poor insulation ratings , bad bonding and lightning protection systems and improperly routing wiring are the most common fire sources

- More than half of the fires reported during the last five years were found to have been caused by wiring and electrical problems, with wire chafing found as the most common problem

- Chafing is one of those problems of scale. The bigger the boat, the more wires there are, and the greater potential there is for chafing issues

Wiring a Fire Suppression System

- This systems are available in 120/240V alternating current system in addition to 24V and 12V direct current systems.

- Electrical connections to bus bars, switches, circuit breakers and other electrical components should be checked for tightness and corrosion on an ongoing basis.

Sources of Power

- Direct current sources for fire protection systems are recommended instead of AC sources because they can be powered all the time

- A constant source of power protected with a proper fuse from the house battery bank is acceptable

Wiring a Fire Suppression System

- Boat vibration and the expansion and contraction of conductors due to temperature changes (thermal cycling) can still cause connections to become loose and/or corroded

- Loose and corroded connections can cause electrical resistance that can lead to heat buildup in conductors, which can cause wire insulation to melt or burn and potentially start a fire

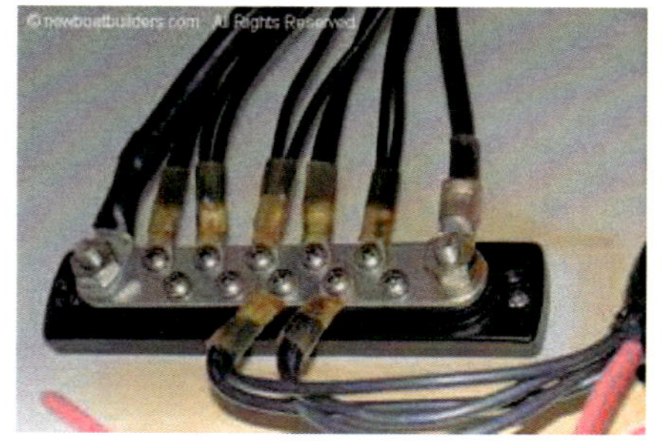

Wiring a Fire Suppression System

- Boat U.S. estimates that 5 percent of boat fires are caused by problems in shore power inlets. Corroded or loose shower power connections can create increased resistance.

- Any blackened or burned shore cord ends should be replaced, and the cause should be investigated.

Wiring a Fire Suppression System

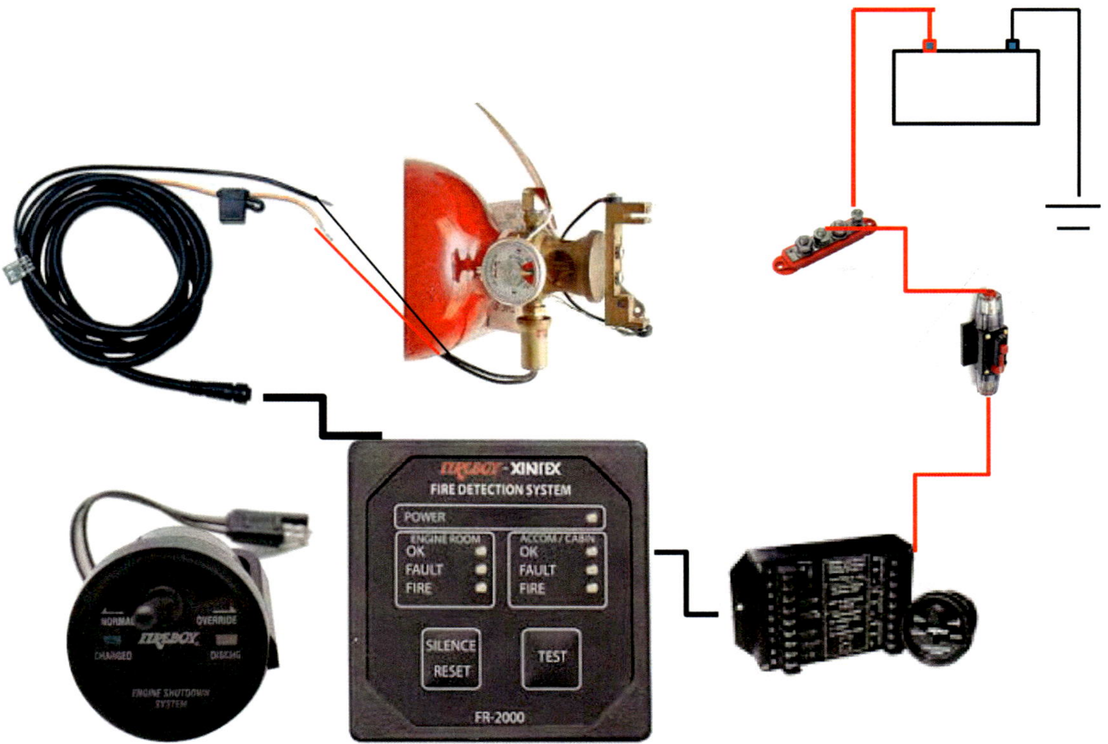

Chapter 10
Air Conditioning and Refrigeration Systems

- Heat Transfer Concepts
- Mechanical ventilation
- Refrigeration Process
- The Evaporator
- The Compressor
- The Condenser
- The Expansion Valve
- Te Capillary Tube
- Water Pump & Pump Relay
- Refrigerants
- Leak Detectors
- Refrigerant Recovery process
- Refrigerant Piping
- The Thermostat
- Ducts & Grills
- Air Conditioner Sizing

Air conditioning

- **Air conditioning** is the cooling of indoor air for <u>thermal comfort</u>

- An air conditioner is an <u>appliance</u>, <u>system</u>, or <u>machine</u> designed to stabilize the air temperature and humidity within an area typically using a <u>refrigeration cycle</u> but sometimes using <u>evaporation</u>, commonly for comfort cooling in buildings , vessels and motor vehicles

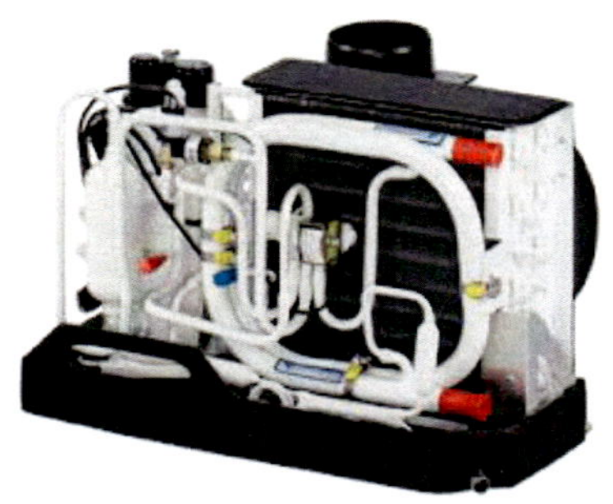

Heat transfer

- **Heat transfer**, is the movement of heat from one place to another
- When an object is at a different <u>temperature</u> from its surroundings, heat transfer occurs so that the body and the surroundings reach the same temperature at <u>thermal equilibrium</u>

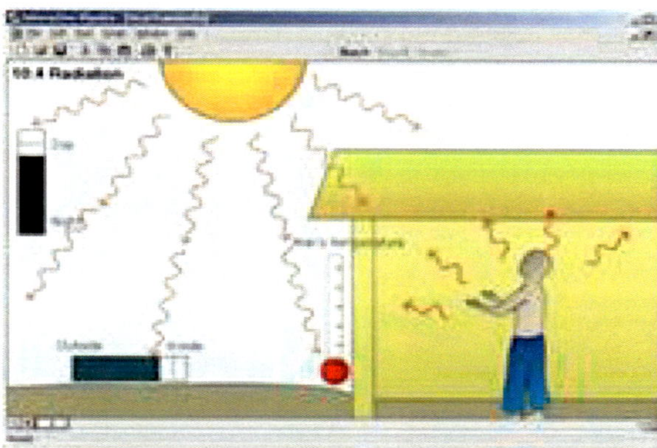

Heat transfer

Heat transfer occurs by convection, radiation or evaporation

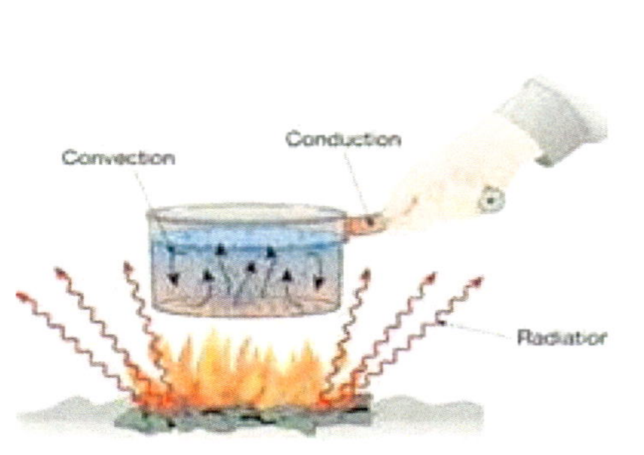

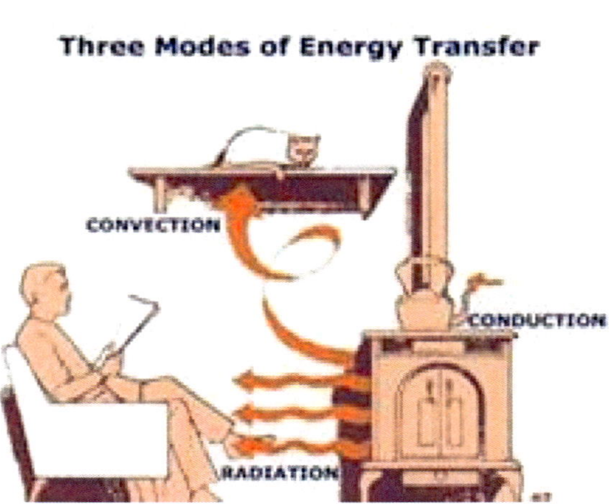

Convection

- Is the transfer of heat from one place to another by the movement of fluids
- When a body gives off heat, the surrounding air becomes warmer and moves upward. Air than contains less heat takes its place

Convection

Two principles of convection are:
- Heat rises (notice how steam or smoke rises)
- Heat always flows from hot to colder

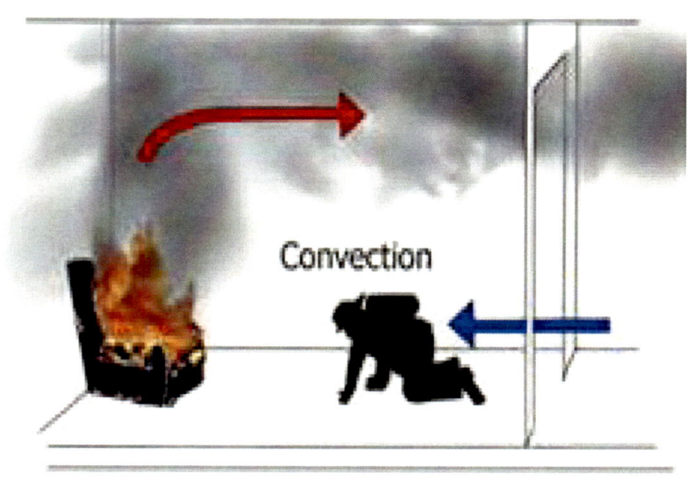

Radiation

- <u>Thermal radiation</u> is the transfer of heat energy through empty space by <u>electromagnetic waves</u>

- The effect's of the sun's radiation are felt by the human body when moving from the shade into bright sun-light

- Dark colors absorb and radiate heat better than light colors

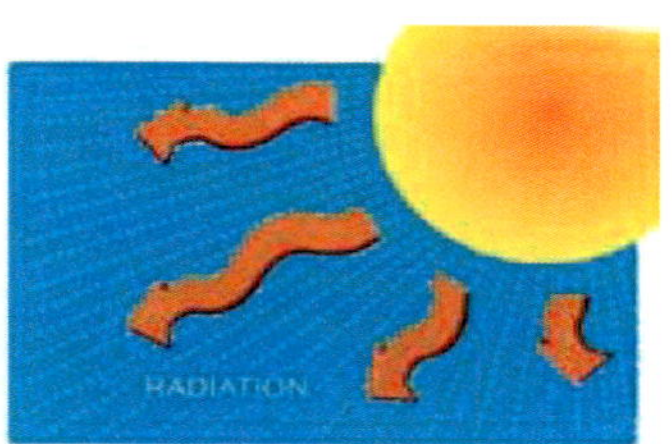

Evaporation

- As moisture is vaporized it absorbs heat, cooling the surface

- Your body feels the effect of evaporation as you perspire

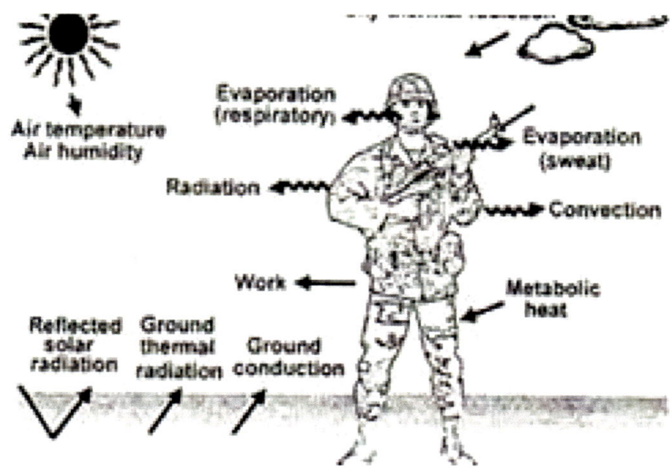

Evaporation

- The cooling effect is more noticeable with wind, as it speeds up the vaporization process

- Evaporation is a type of phase transition; it is the process by which molecules in a liquid state (e.g. water) spontaneously become gaseous (e.g. water vapor)

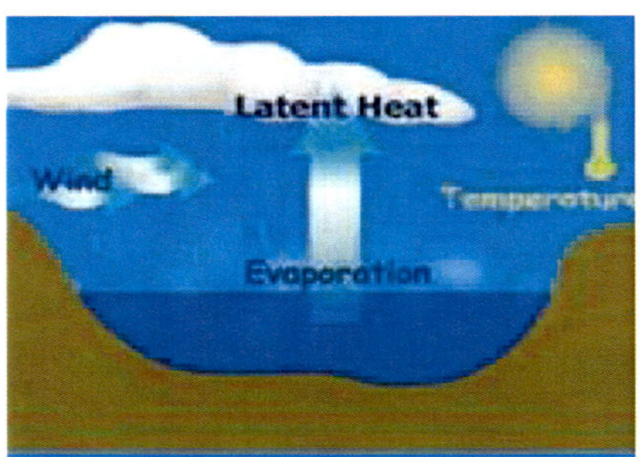

Temperature & Humidity

- Our perception of temperature is due to the transfer of heat between our bodies and the environment

- Temperature is a measure of our perception of hotness or coldness. The faster we lose heat to a substance, the colder it feels to us

- Besides cooling the air, air-conditioning system provide humidity control too

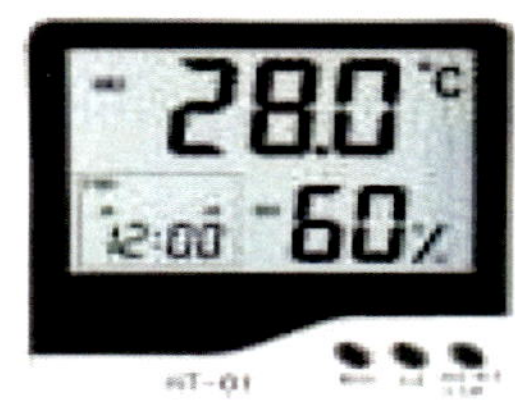

Temperature & Humidity

- When humidity is 100% the air is totally saturated with moisture.
- When humidity is 50% , the air is holding half the amount of moisture that it is capable of holding at a given temp.
- Low Humidity (dry air) permits heat to be taken away from the human body by evaporation of perspiration.
  - After a shower if a breeze hits you, you feel cool from evaporation

States of Matter

Solid is the state in which matter maintains a fixed volume and shape; liquid is the state in which matter maintains a fixed volume but adapts to the shape of its container; and gas is the state in which matter expands to occupy whatever volume is available

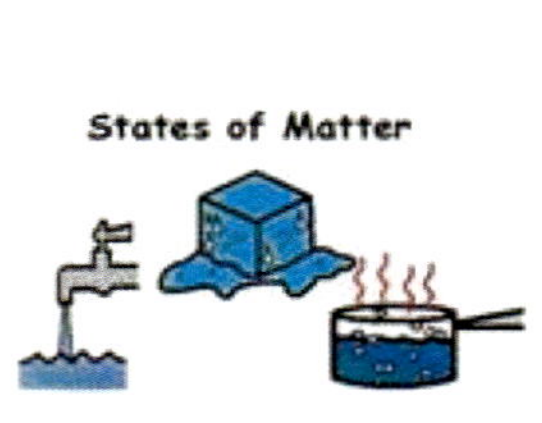

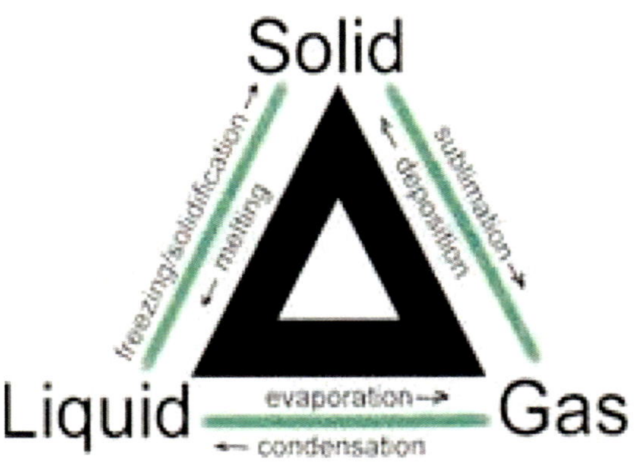

Latent Heat

- The expression **latent heat** refers to the amount of energy released or absorbed by a chemical substance during a change of state that occurs without changing its temperature

- The **specific latent heat** is the amount of energy required to convert 1 kg (or 1 lb) of a substance from solid to liquid (or vice-versa)

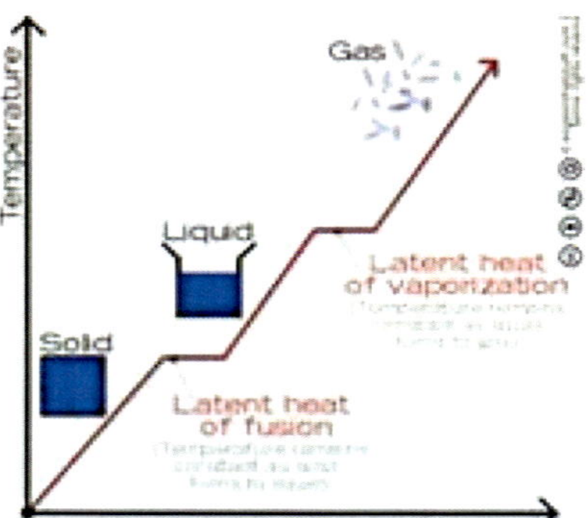

Latent Heat

- After the refrigerant reaches its boiling temp.(-22 F @ atmospheric pressure) it absorbs more and more heat, yet its temperature does not increase. It is simply absorbing heat as it attempts to change its state from a liquid to a gas

- For this reason "Latent heat" cannot be recorded on a thermometer

Mechanical or forced ventilation

- It is used to control indoor air quality
- Excess humidity, odors, and contaminants can often be controlled via dilution or replacement with outside air. However, in humid climates much energy is required to remove excess moisture from ventilation air

Mechanical or forced ventilation

Ceiling fans and table/floor fans circulate air within a room for the purpose of reducing the perceived temperature because of evaporation of perspiration on the skin of the occupants

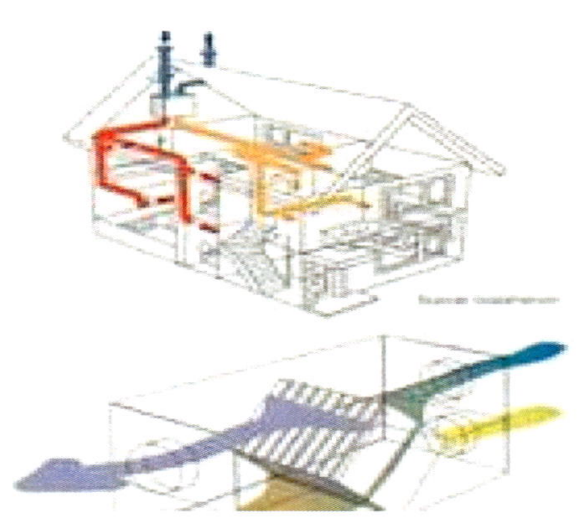

Refrigeration Cycle

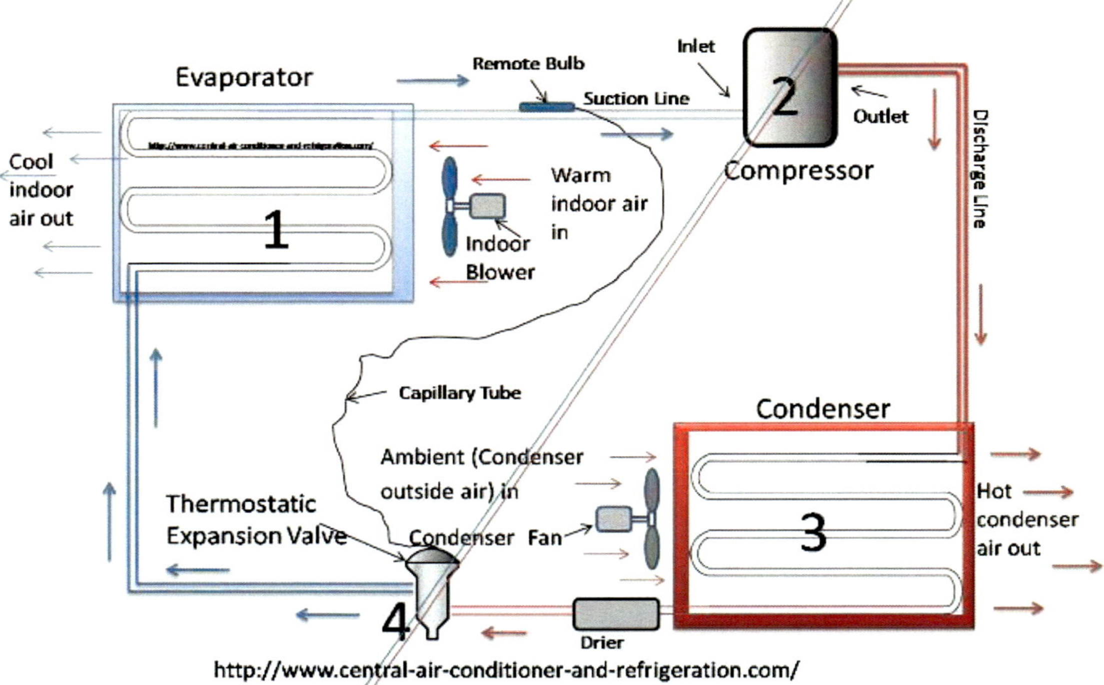

Five basic components

- The compressor.

- The condenser.

- The expansion device.

- The evaporator

- The copper refrigerant tube (a tube that connects these air conditioner parts)

Refrigeration Process

In this refrigeration diagram, the four major components split into two sections: Indoor and Outdoor. In indoor units, we have the Evaporator coil and the blower. In outdoor units, we have the Compressor and the Condenser

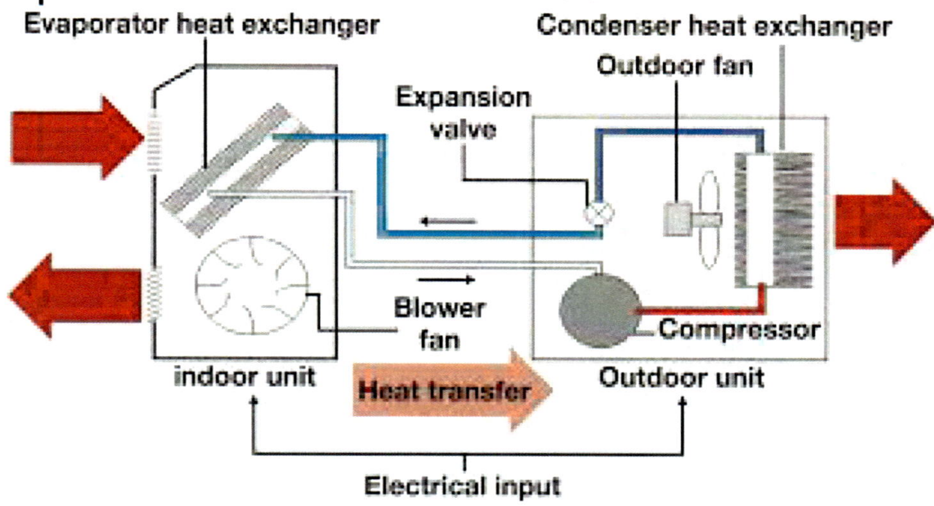

High Pressure & Low Pressure

- The high pressure side is the condenser units (outdoor)
- The low pressure side is the air conditioning evaporator (indoor).
- The divided point between high and low pressure cut through the compressor and the expansion valve
- Refrigeration cycle is a process that removes heat from indoor evaporator to outdoor condenser units

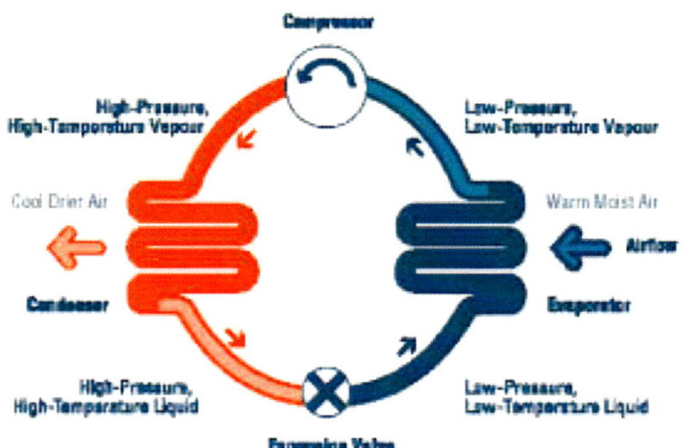

Residential A/C

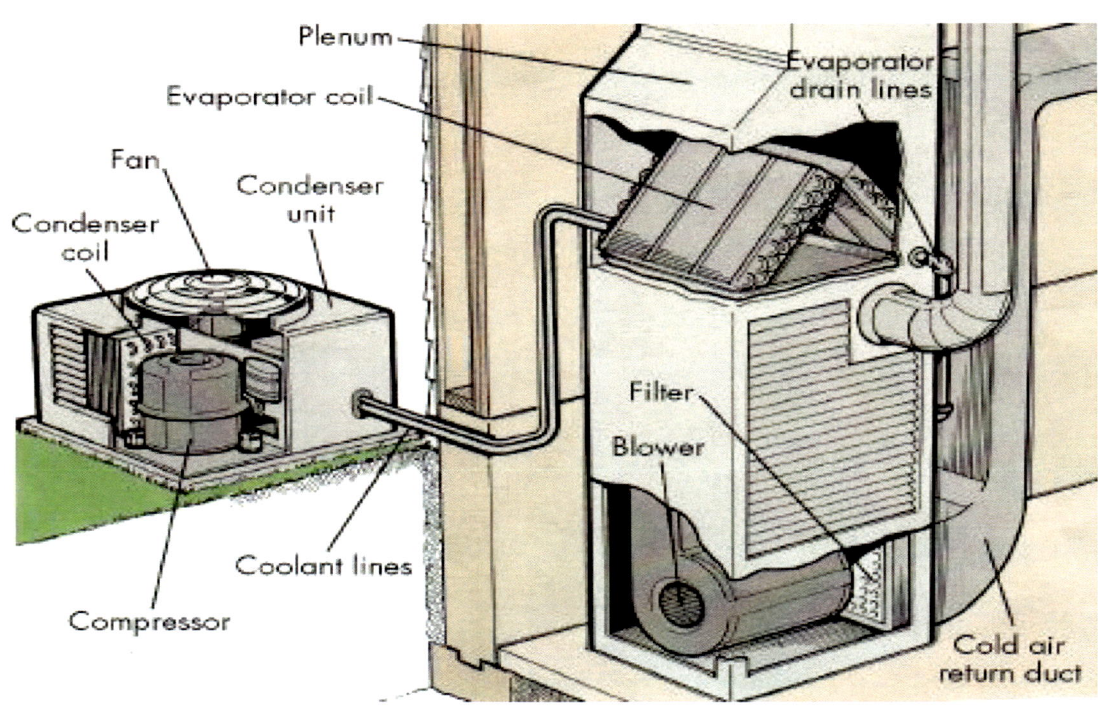

A/C Residential

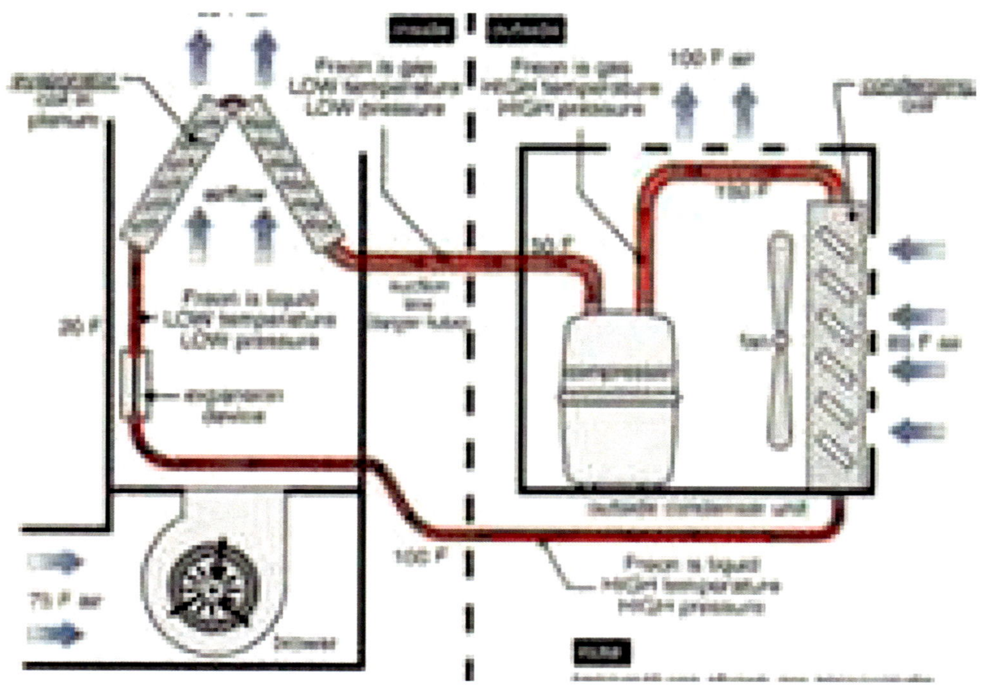

Residential A/C

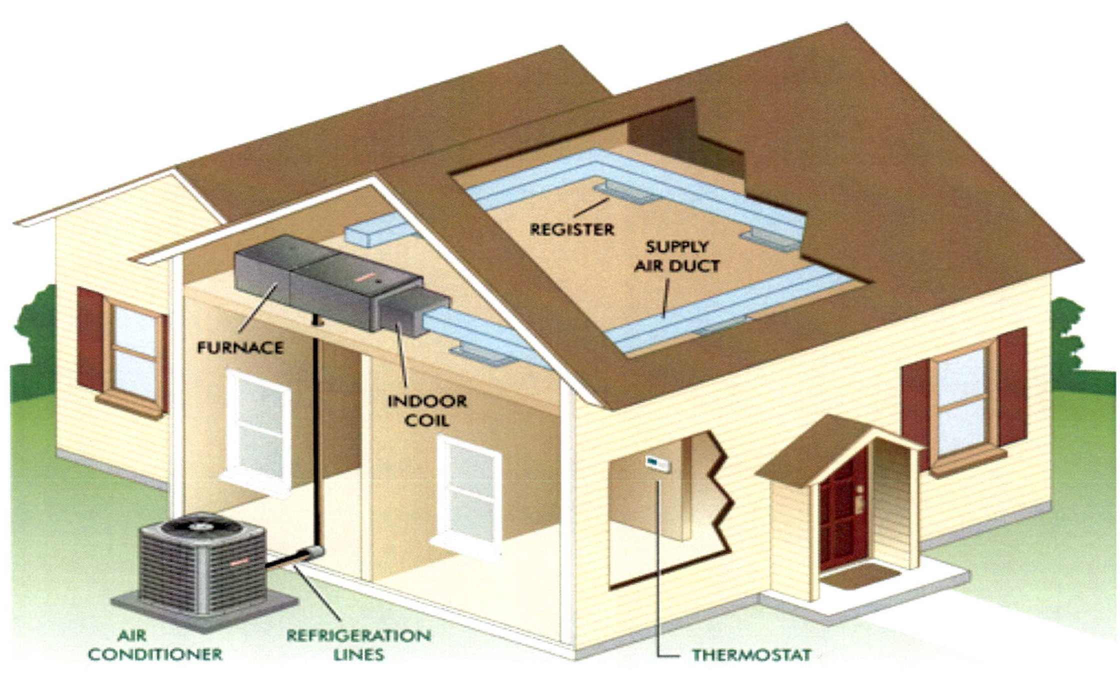

Automobile A/C

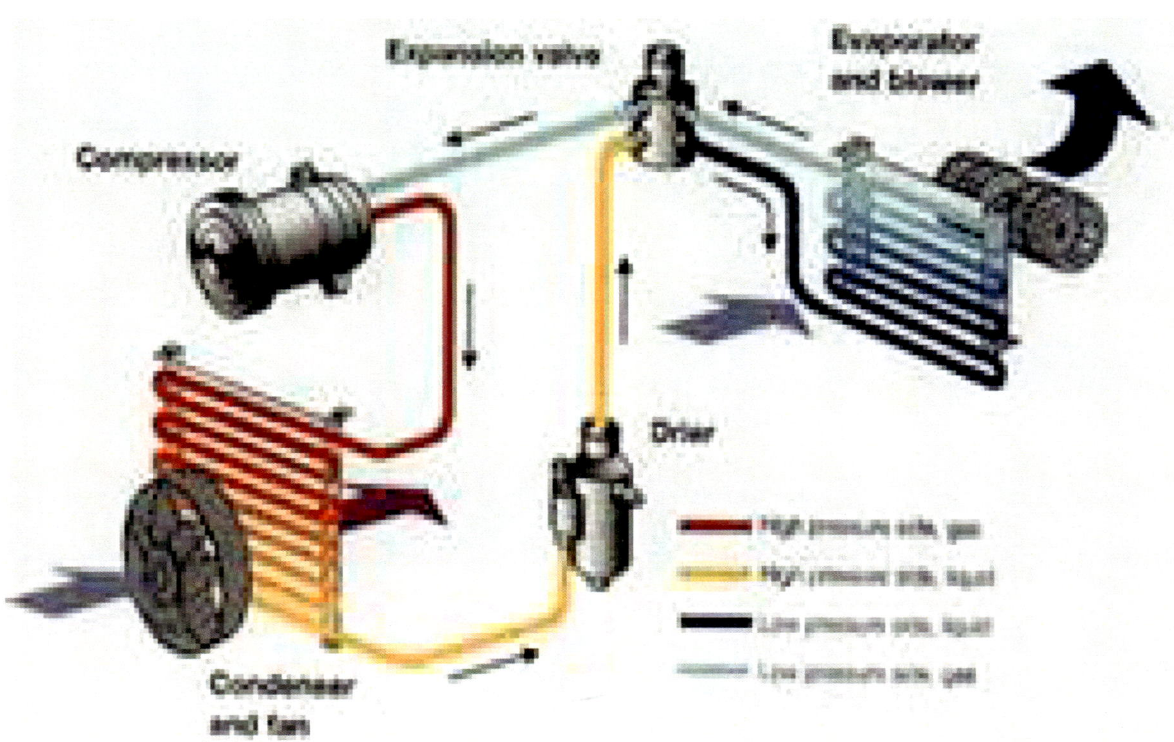

Automobile A/C

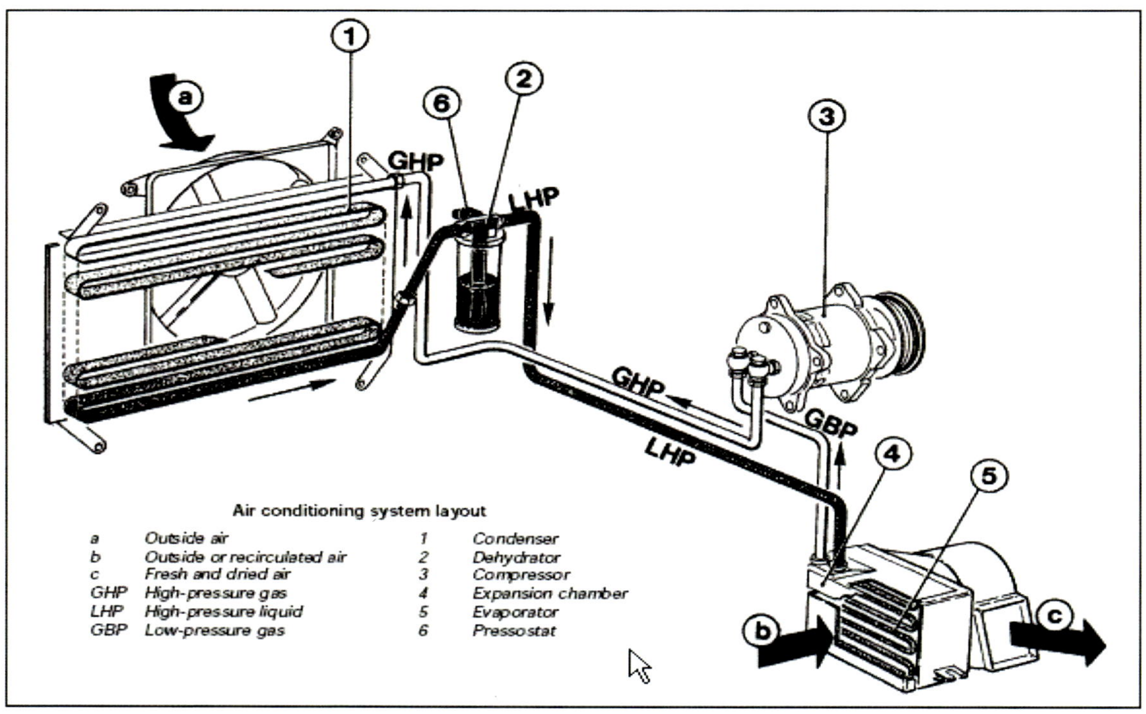

Air conditioning

- Air conditioning and refrigeration are provided through the removal of heat.

- The definition of cold is the absence of heat and all air conditioning systems work on this basic principle.

- An air conditioning system, or a standalone air conditioner, provides cooling, ventilation, and humidity control for all or part of a house or building.

Evaporator

Evaporator

The air conditioning evaporator is a heat exchanger that absorbs heat into the air conditioner system. The evaporator does not exactly absorb heat! It's the cooled refrigerant fed from the bottom of the evaporator coils absorb the heat

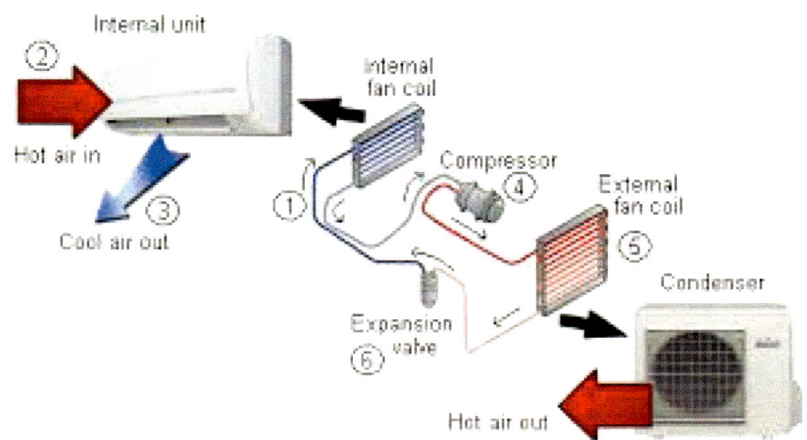

Evaporator

- The liquid refrigerant usually flows from the bottom of the evaporator coils and boils as it moves to the top of the evaporator coils
- The reason it's fed from the bottom is to ensure the liquid refrigerant boils before it leave the evaporator coils

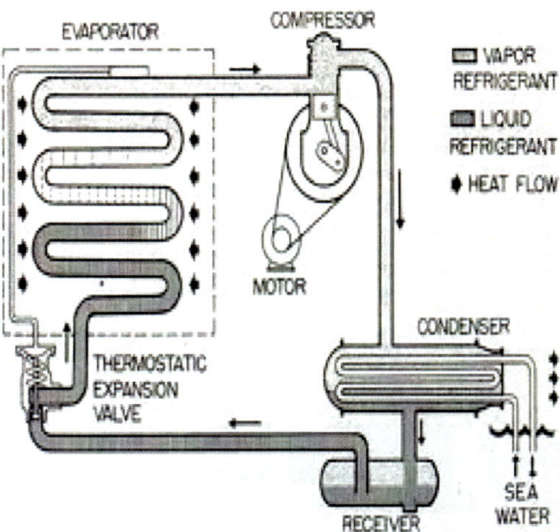

Evaporator

- If a refrigerant was to fed from the top, the liquid refrigerant would easily drop to the bottom of the coils before it absorbs enough heat and boil
- If evaporator was too feed liquid refrigerant into air conditioner compressor; it will shorts the air conditioner compressor life
- The air conditioner evaporator has three important tasks:
  - Its absorb heat
  - Boils all the refrigerant to saturated vapor
  - Superheat

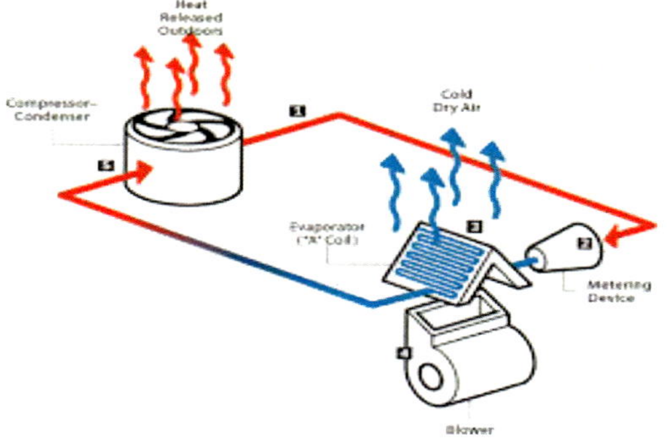

254

Evaporator (Remote Fan Coil)

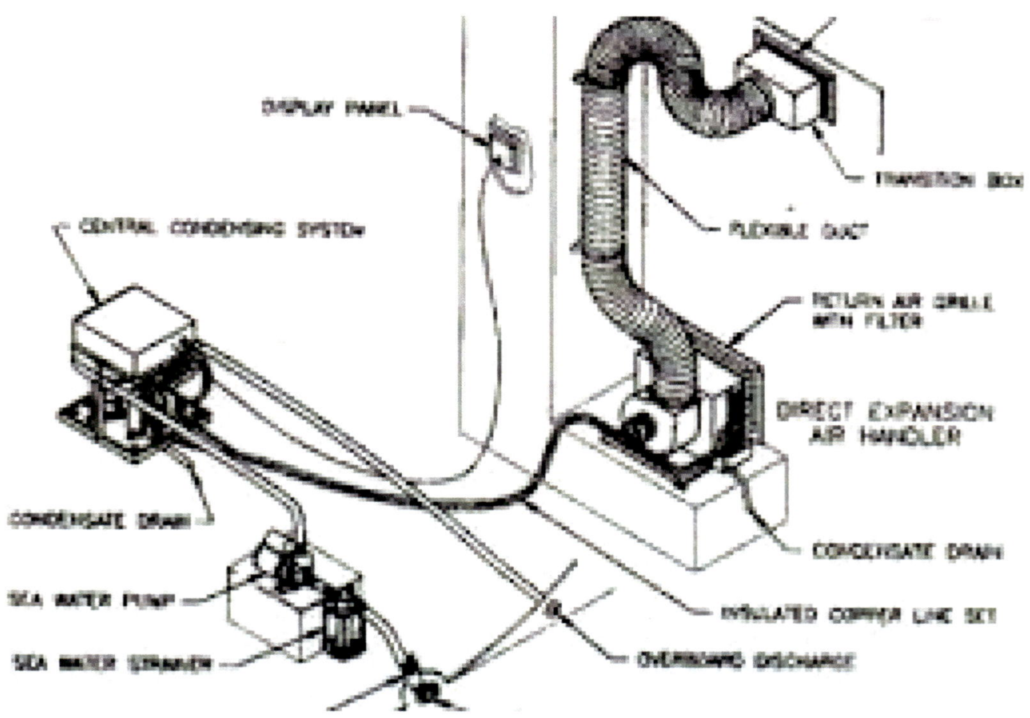

Remote Fan Coil

- Remote fan coils or **split systems** or remote Evaporator are used in boat applications
- The evaporator coil is connected to a remote condenser unit using piping instead of ducts
- Dehumidification in an air conditioning system is provided by the evaporator. Since the evaporator operates at a temperature below dew point, moisture in the air condenses on the evaporator coil tubes. This moisture is collected at the bottom of the evaporator in a condensate pan and is removed by piping it to a central drain or onto the Boat outside

255

Remote Fan Coil

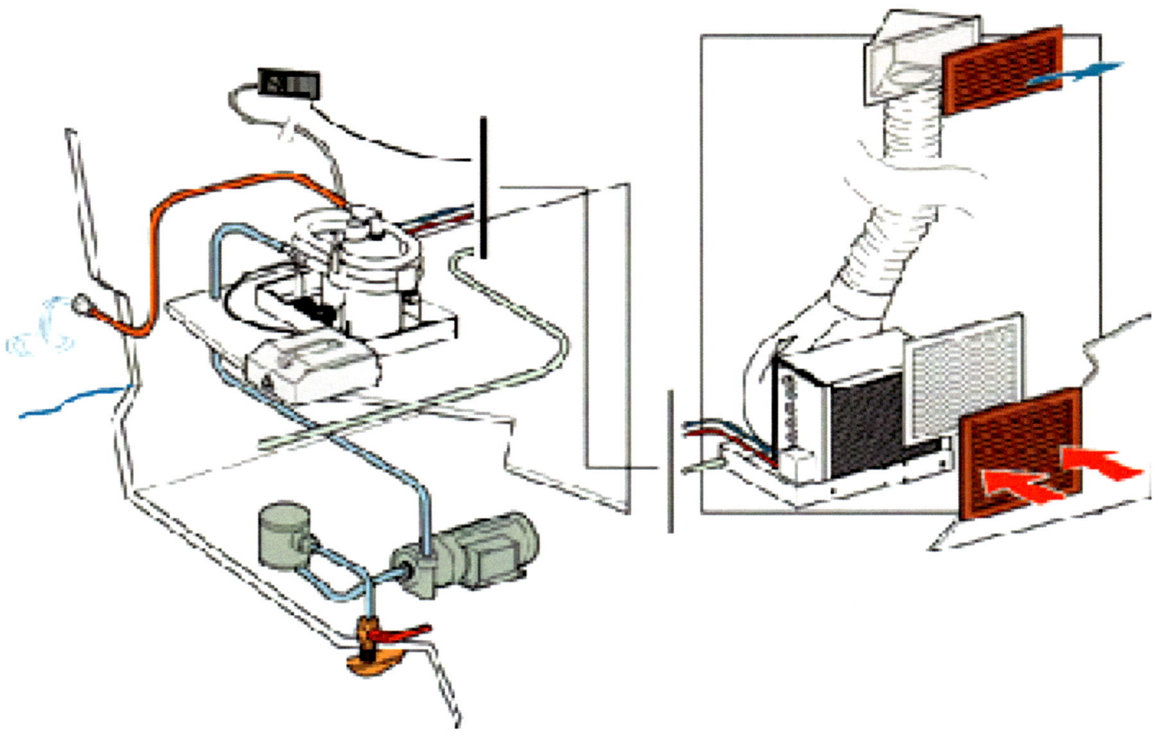

How the Evaporator work?

- A (40°F) refrigerant flows through the evaporator, it absorbs 75°F indoor heat, causing the liquid refrigerant in the evaporator to boils.

- the temperature of the refrigerant will always try to equalize. The 75°F heat will flow to 40°F refrigerant and it will increases the 40°F temperature and boils it.

- After the liquid refrigerant travel across the evaporator coils, the entire liquid refrigerant should have boils. At this point it's known as saturated vapor point

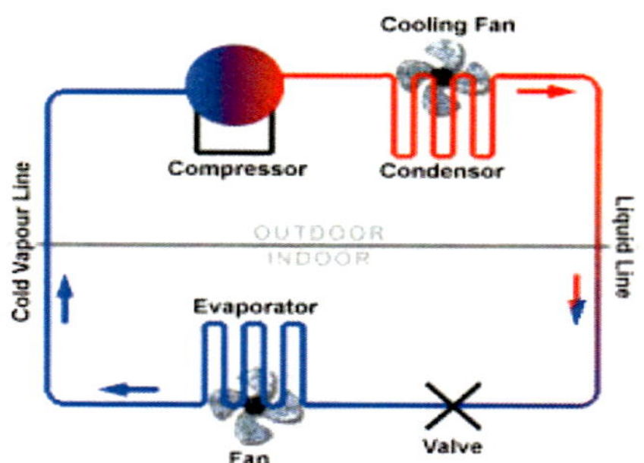

Reducing Humidity

- When moisture condenses on the cool window surfaces, visibility problems can result.

- When warm air passes across the cool fins of the evaporator, it loses heat.

- Moisture in the air tends to condense on the evaporator fins, like it does on the cool glass of a mirror.

- The moisture that deposits on the cold evaporator core collects in a drain pan

The Condenser Drain Line

Dehumidification in an air conditioning system is provided by the evaporator. Since the evaporator operates at a temperature below dew point, moisture in the air condenses on the evaporator coil tubes. This moisture is collected at the bottom of the evaporator in a condensate pan and is removed by piping it to a central drain or onto the Boat outside

The condensate drain line

- The condensate drain line should never be connected to the raw water discharge line.
- if there were a blockage in the discharge line, the raw water pump would pump raw water back into the drain pan and ultimately into the boat causing a sinking hazard.
- The condensate drain line is a gravity line and the highest point of this line will determine the water level in the condensate pan. A common mistake is to have the condensate line run a couple of feet at the same level of the chassis

Condensate Drain

- Condensate drains from evaporator systems shall not terminate within three feet of the outlet of engine exhaust systems on the exterior of the hull.
- Condensate drains must be oriented so that they are routed in a continuous down hill direction.
- It is acceptable to install a "p-trap" in this line to contain a small amount of water, in order to prevent CO migration up.

The compressor

- The refrigerant cycle consists of four essential elements to create a cooling effect
- **<u>A compressor</u>** provides compression for the system

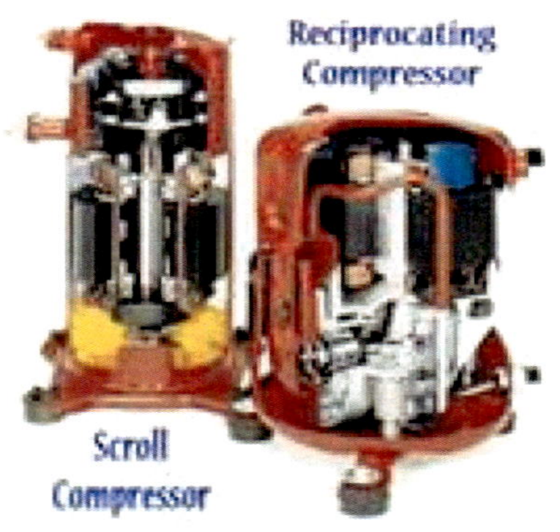

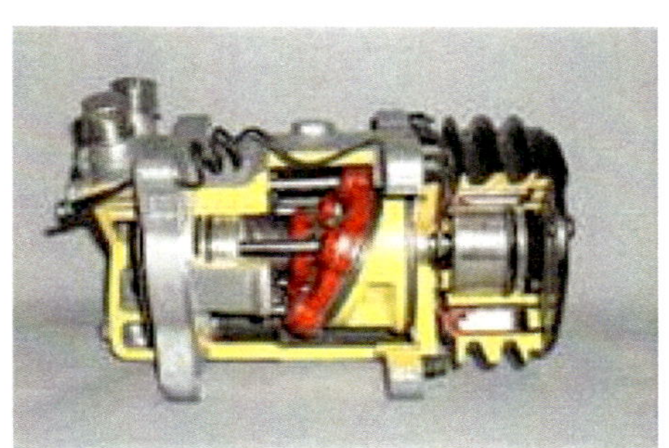

Compressor

- The air conditioning compressor is known as the heart of the air conditioner units
- It's one of the divided points between high and low side.
- The compressor inlet lines are known as:
  - Suction pressure
  - Back pressure
  - Low side pressure
- The compressor outlet lines are known as:
  - High side pressure
  - Discharge pressure
  - Head pressure

How the Compressor work?

The compressor absorbs vapor refrigerant from the suction line and compresses that heat to high superheat vapor

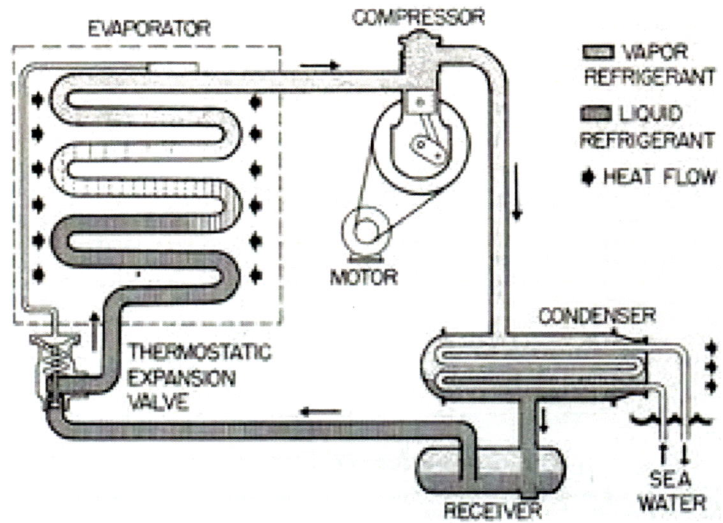

How the Compressor work?

- As the refrigerant flows across the compressor, it also removes heat of compression, motor winding heat, mechanical friction, and other heat absorbs in the suction line

- The air conditioner units compressor produce the pressure different, it's the air conditioner compressors that cause the refrigerant to flow in a cycle

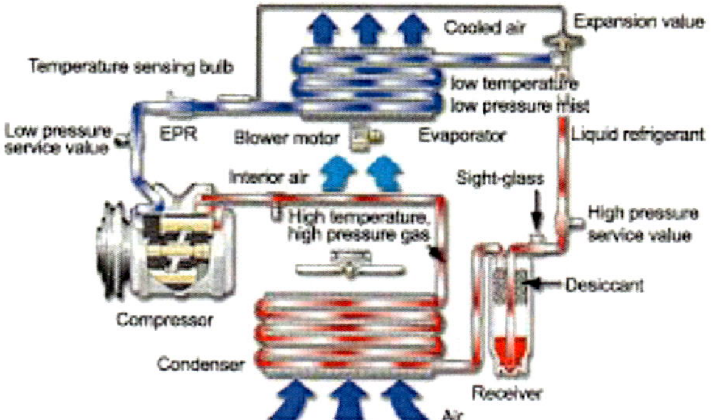

The compressor

This compression causes the vaporized cooling medium to become more dense; a by-product of this process is added heat

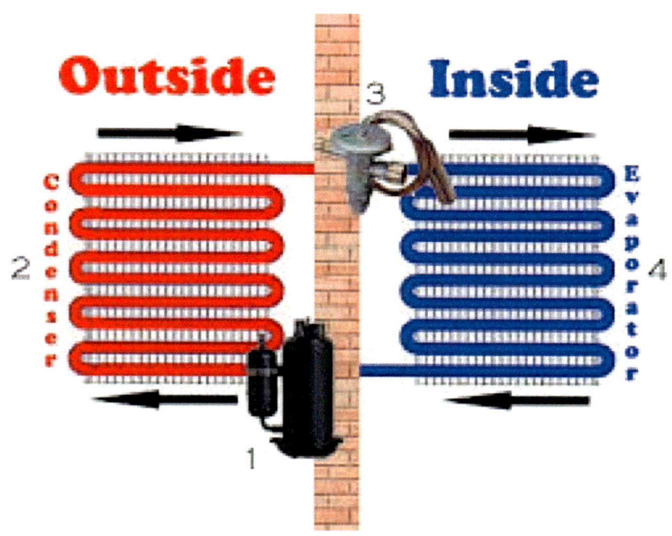

The refrigerant Flow

Refrigerant is drawn from the evaporator (4) and pumped to the condenser (2) by the compressor (1). The compressor also pressurizes the refrigerant vapor so that it will change state (condense) readily

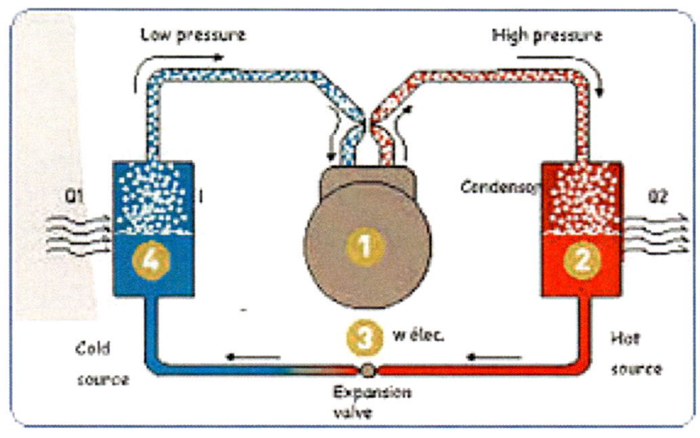

Compressor

A reciprocating compressor with 40 mm cylinder length, is able to deliver about 150 PSI of pressure

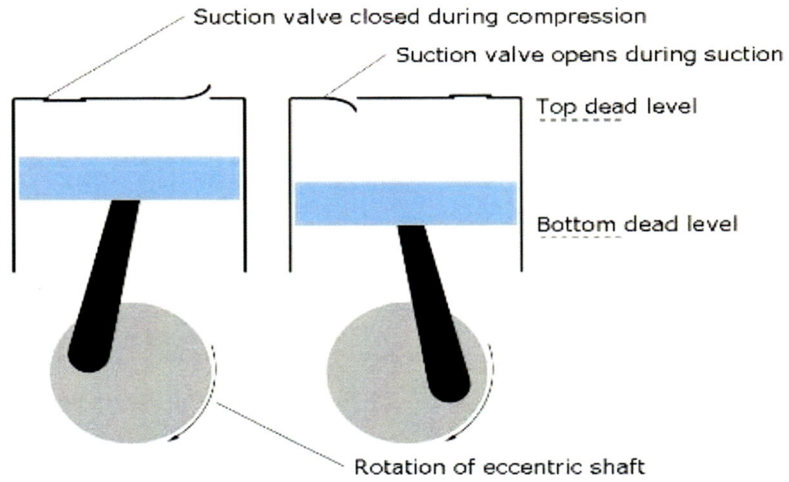

air-conditioner-selection.com. Reproduction with permission only

The Condenser

Cutaway view of *Bryant* condenser unit

1. Circuit board and switches
2. Copper tube
3. Fan and fan cover
4. Compressor
5. High and low pressure switches
6. Condenser tube and fins
7. Muffler
8. Liquid line service valve
9. Suction line service valve
10. Condenser casing

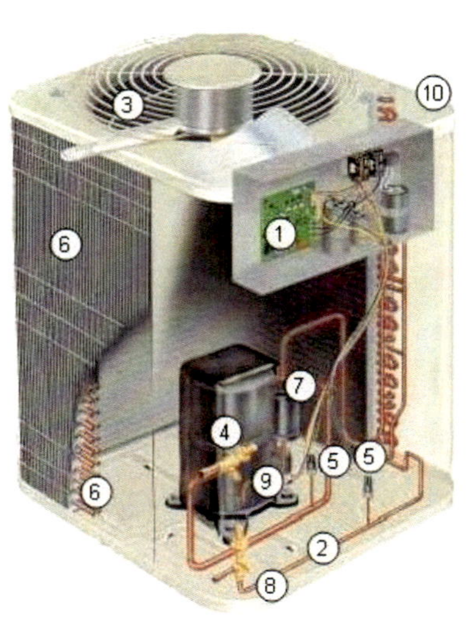

Condenser

Under normal operating conditions, the condenser (the copper tube within a tube on the unit) should be warm to the touch - if you cannot feel any heat in the condenser whatsoever when the compressor is running, you have too much cold water passing through it

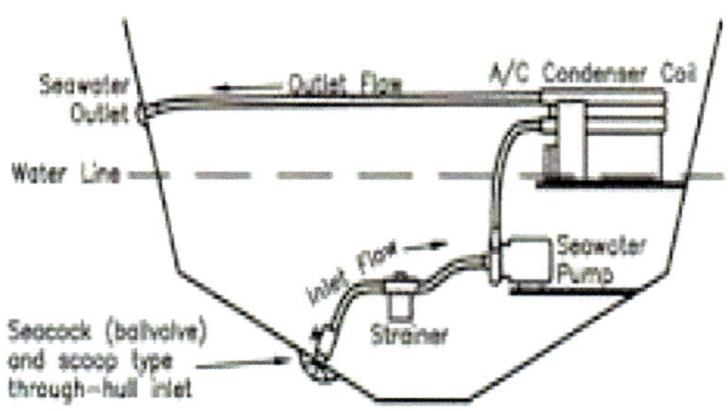

Emerald Small Condenser

Condenser Components

- The Condenser unit is composed by : The Compressor, The Accumulator and the Heat Exchanger Coil

The Condenser Unit

The Condenser Coil (5) is refrigerated by means of the sea water coming from the the seacock (2) and discharge the warm water directly outboard (8)

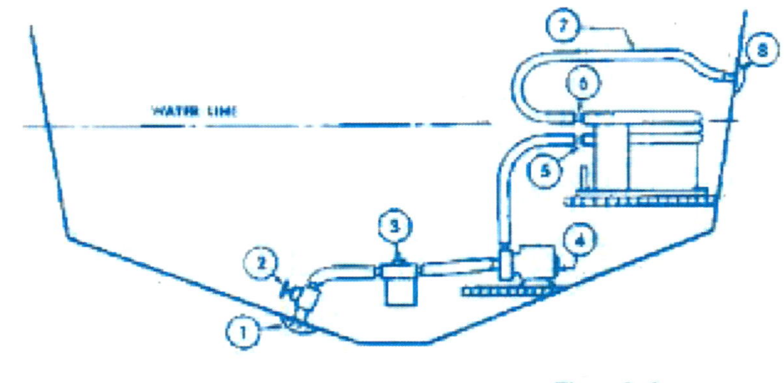

Figure 3 - 3

One Condenser Two Evaporators

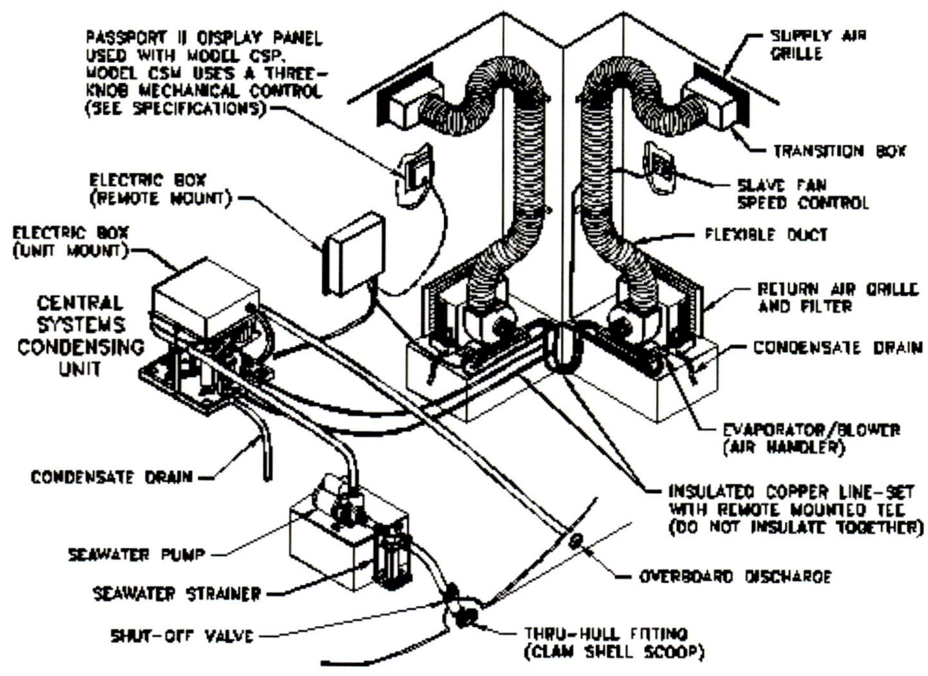

The compressor operation

- The compressor keeps the refrigerant flowing through the system at specific rates flow and pressure

- The low pressure switch is on the suction side of the compressor

- A drop in suction pressure, for whatever reason will shut the system down

High Pressure Switch

The high pressure switch would be located on the high pressure side and its purpose is to protect the compressor

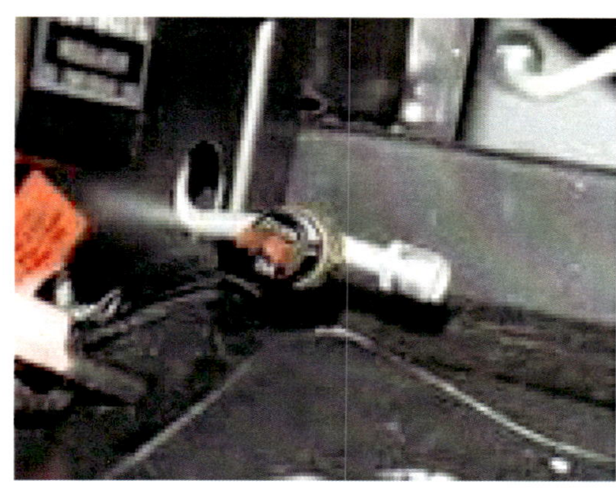

The process

- The compressor takes refrigerant vapor in form the low pressure side of the circuit and discharges it at a much higher pressure into the high side of the circuit
- Most air cooled air conditioning systems are designed so that the refrigerant will condense at a temperature about 25 to 30 degrees above outside ambient air temp

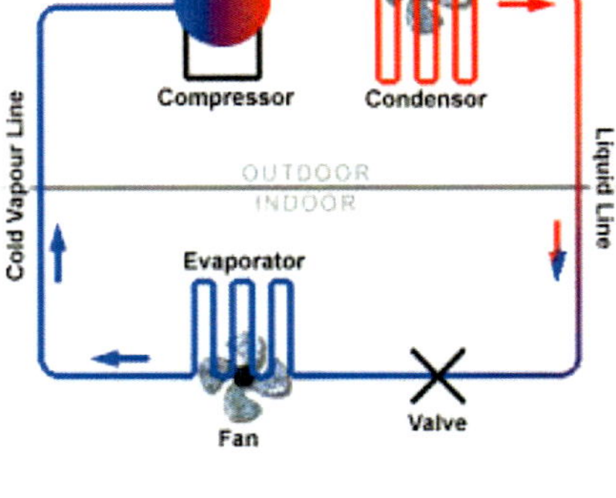

Marine applications

- When the hot refrigerant vapor discharged from the compressor travels through the condenser, the cool air flowing through the condenser coil absorbs enough heat from the vapor to cause it to condense
- If the outside Temp. Is 80 degrees, the system is designed so that the temperature of the refrigerant, right at the point where it first condenses, will be about 105 to 115 degrees

The Refrigeration Cycle

Refrigerant enters the compressor in the form of a gas, where it is compressed by the compressor to a higher pressure and temperature state

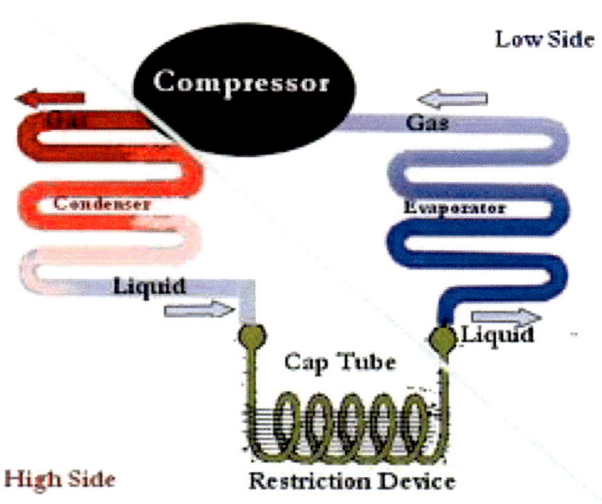

The refrigeration Cycle

The gas enters the condenser. where precipitation and heat converts the gas to a high-temperature liquid, which then flows through the AC expansion valve

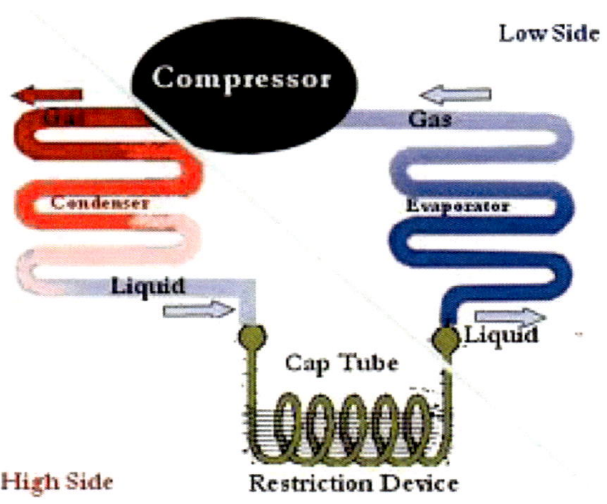

The Refrigeration Cycle

The expansion valve allows a regulated amount of fluid to enter the evaporator, which keeps pressure low and allows the liquid to expand back into a gas state

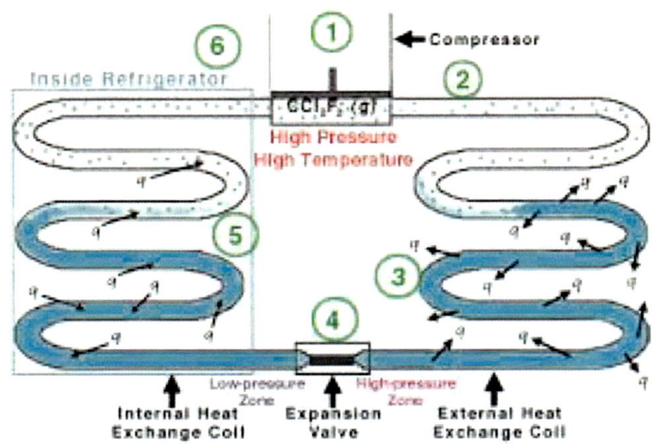

Expansion Valve

A metering device (often called "Expansion valve") acts as a restriction in the system at the evaporator to ensure that the refrigerant flows into the evaporator at the proper rate; this keeps the refrigerant from returning to the compressor in a liquid state, and also controls the rate of heat exchange in the evaporator.

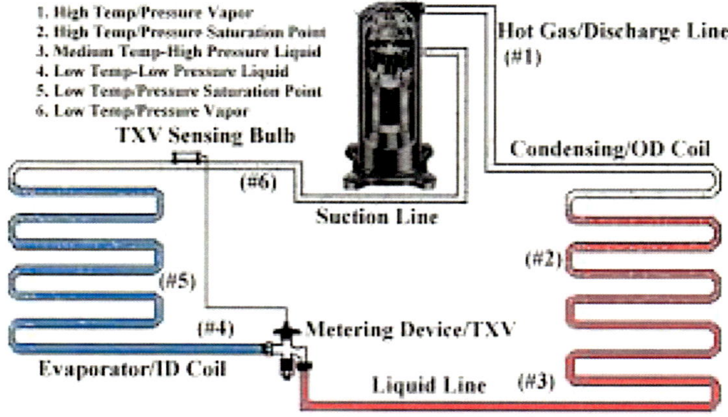

Metering Device

- It is the dividing point between the high pressure and low pressure sides of the systems

- When the refrigerant passes through to the expansion valve, it drops from about 225 psi to 70 psi. It also drops in temperature from about 110 degrees to about 40 degrees, it starts evaporating immediately

Types of expansion valve

- There are two types of ac expansion valve in air conditioner units:
  - Thermostatic expansion valve (TX)
  - Capillary tubes

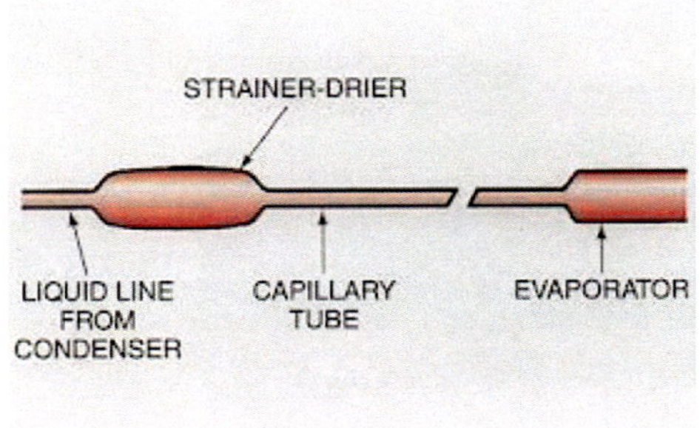

How does air conditioner expansion valve works?

AC expansion valves work by controlling the amount of refrigerant flows to the evaporator coils. It acts as restriction to provide a specific amount of refrigerant flows into the evaporator coils

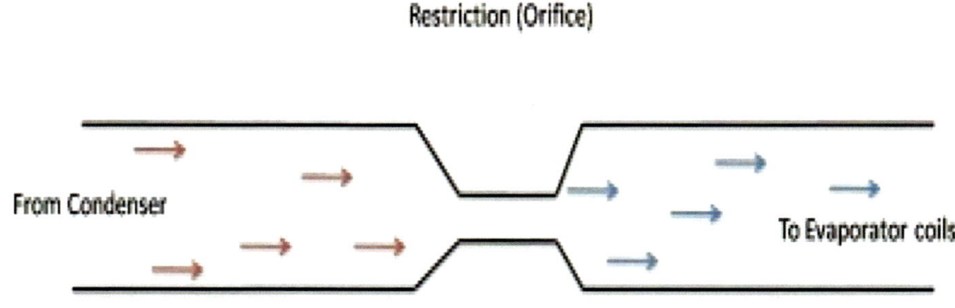

The Venturi Principle

The Venturi effect is a jet effect; as with a funnel the velocity of the fluid increases as the cross sectional area decreases, with the static pressure correspondingly decreasing

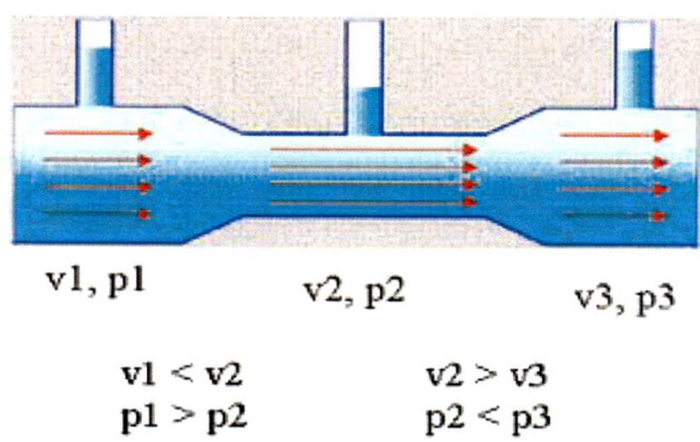

The ventury Principle

When a fluid passes through the tube's throat (narrow section) it moves faster. When the fluid moved faster the pressure decreased. That means the greater the speed of the air the les pressure exerted

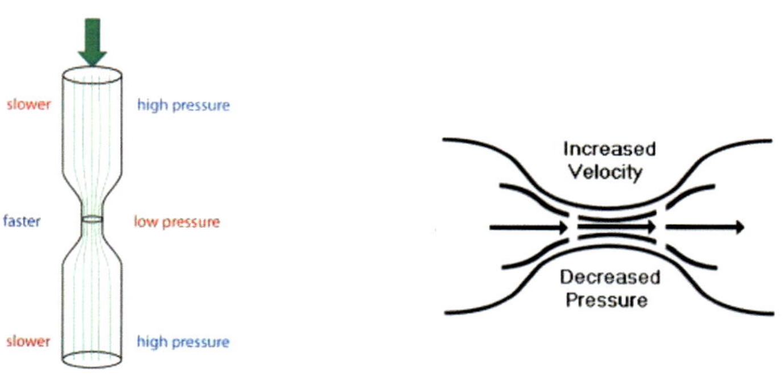

Metering Device functions

- **First:** It controls the amount of liquid refrigerant entering the evaporator coils. The amount of liquid refrigerant entering the evaporator must equal the amount of refrigerant boils in the evaporator coils

- **Second:** It maintains a pressure difference between the high and low pressure sides of the system to permit the refrigeration to vaporize

Thermal expansion valve

Thermal expansion valves are often referred to generically as "metering devices

Thermal expansion valve

A **thermal expansion valve** (often abbreviated as TXV or TX valve) is a component in refrigeration and air conditioning systems that controls the amount of refrigerant flow into the evaporator thereby controlling the superheat at the outlet of the evaporator

Expansion Valve (EV)

(Capillary tube, Expansion valve) The metering device restricts the flow of liquid refrigerant from the condenser to the evaporator. As refrigerant passes through the metering device, its pressure decreases

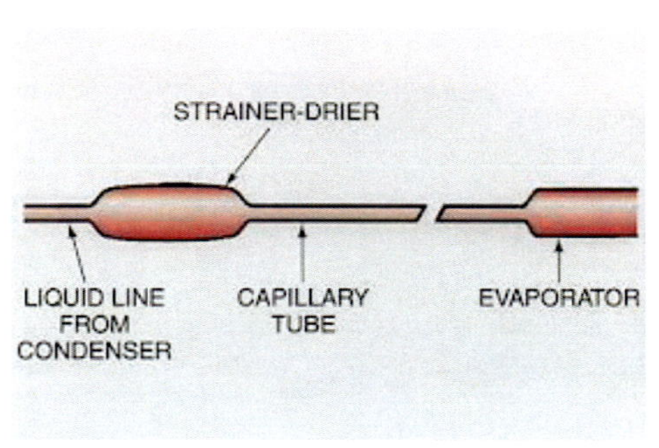

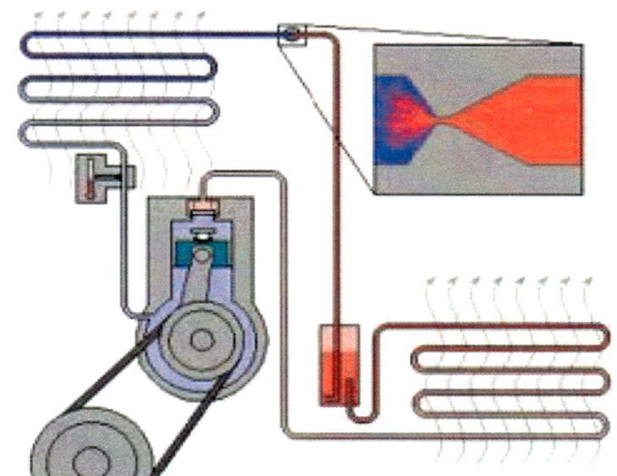

Strainer Drier

The Drier is located before the expansion valve to protect the valve and the capillary tube from moisture and debris

Capillary tube metering device

The capillary is simply a length of tubing with a small inside diameter which acts as a constant throttle on the refrigerant entering the evaporator

Capillary tube metering device

- A fine filter or filter drier installed at the inlet of the capillary prevents dirt from blocking the tube

- The capillary tube equalizes the pressure in the system when the unit stops. This pressure equalizing characteristic of the capillary allows a low starting torque motor to be used with the compressor

How does capillary tube works?

- Capillary tube works by restricting and metering the liquid flow, the capillary tube can maintain the required pressure differential between the condenser and the evaporator

- Because of friction and acceleration, the pressure drops as the liquid flows through the tube

How does capillary tube works?

- In order to reduce the temperature of the liquid to the saturation temperature of the evaporator, some of the liquid must turn into a vapor in the capillary tube, "flash", just as it does with all refrigerant controls

- All air conditioner expansion valve works in similar fashion. It shapes, size, capacity, and manufacture are different, but it operation principle are alike

How does capillary tube works?

- The capillary tube diameter and length must be such that the flow capacity at the design pressures (condensing and evaporating) equals the compressor pumping capacity at these same conditions

- For example, if the tube diameter is too small (resistance to high) the liquid refrigerant flow will be less than the pumping capacity of the compressor with the evaporator being "starved" and the suction pressure being low

How does capillary tube works?

Less liquid will enters the evaporator and the excess will build up in the condenser reducing the effective condensing surface and increasing the condensing temperature and pressure. This pressure change tends to increase the flow in the tube and at the same time reduces compression capacity

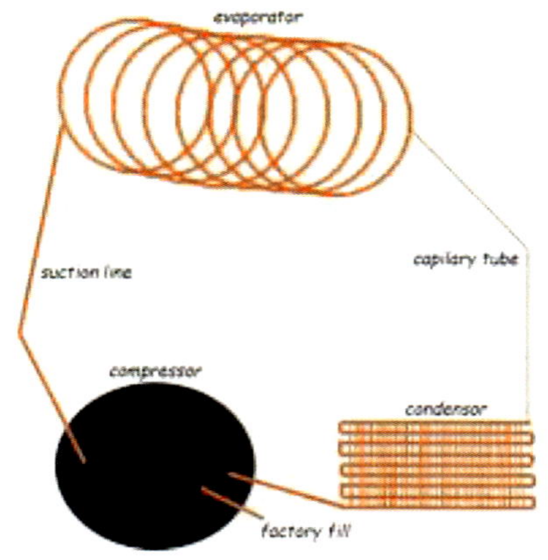

TEV Valve Operation

- AC **T**hermostatic **E**xpansion **V**alves contain a temperature-sensing bulb filled with gas

- This gas forces the expansion valve to close and press against a spring in the <u>body</u> of the valve as bulb temperature increases

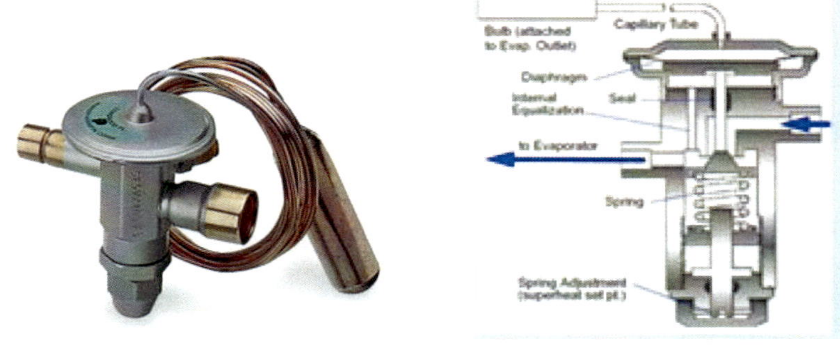

TEV Valve Operation

As evaporator temperature decreases, so does bulb pressure, which relaxes the spring and enables the AC expansion valve to open

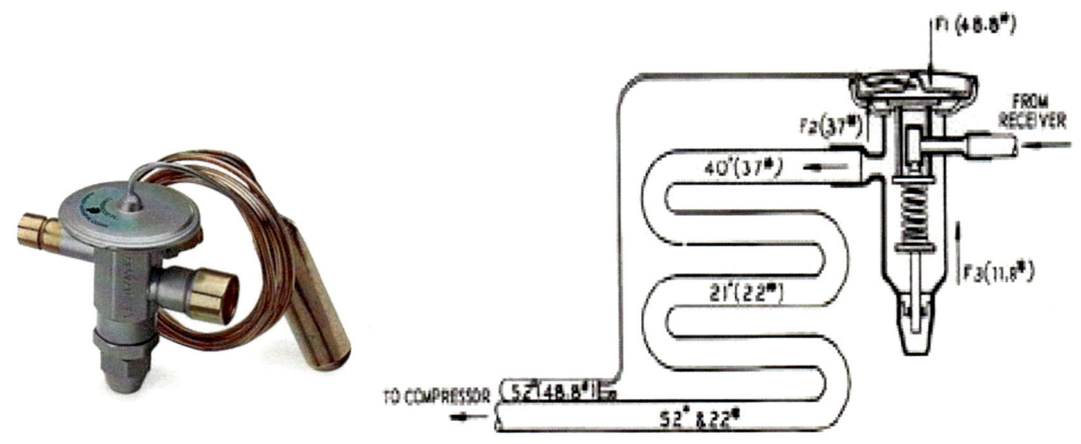

Capillary Tube Vs ETV

While the capillary tube is used in the small domestic systems Pleasure yachts, the thermostatic expansion valve is used in the systems of higher capacities

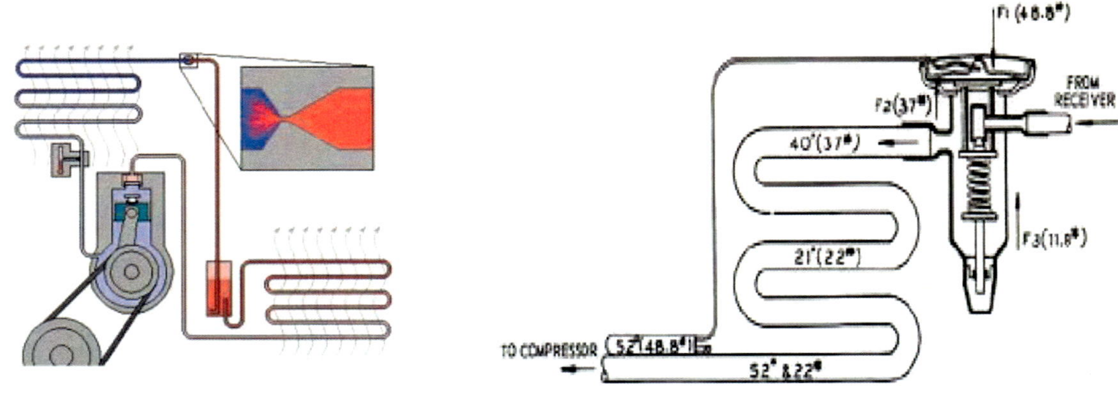

ETV Valve Uses

ETV is commonly used in the industrial refrigeration plants, high capacity split air conditioners, packaged air conditioners, central air conditioners and automotive systems

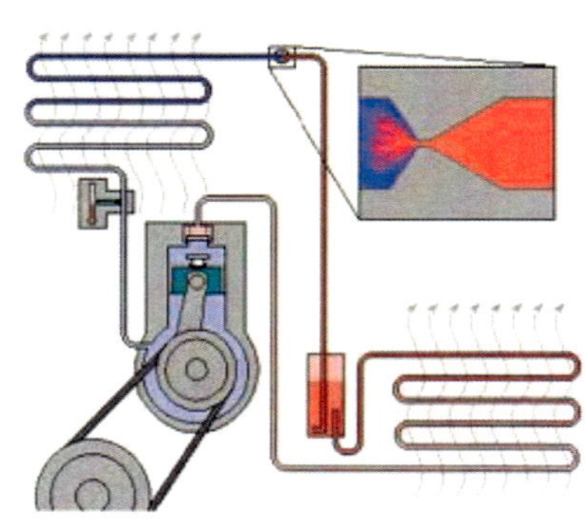
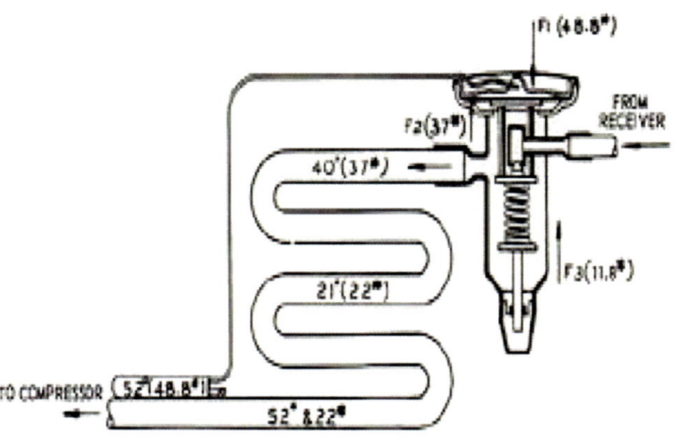

Types of ETV's

- Bulb Sensor.

- Electronic Controlled

- Adjustable Pressure

Water Pump & Plumbing

- The sea water requirements for cooling our equipment is about 200 gallons/hour/ton of air conditioning(12,000BTUs= one ton) at a maximum temperature of 100°F.

- All of our condensing tubes are 5/8"OD, and for systems of up to about three tons of A/C (36,000 BTUs), a single 3/4" intake through hull set up is adequate - larger systems may need a 1" or larger, depending on the total GPH requirements

- The only metal the salt water touches is the inside of a 90/10 CuNi condensing tube which is inside a copper shell

Pump Relay

In multiple unit installations, I often suggest using one large pump with a pump relay box whereby each unit has a 12/24 volt wire returning to the pump relay box (usually mounted by the pump), and if any one of the units engage, the pump will supply water to all of the units.

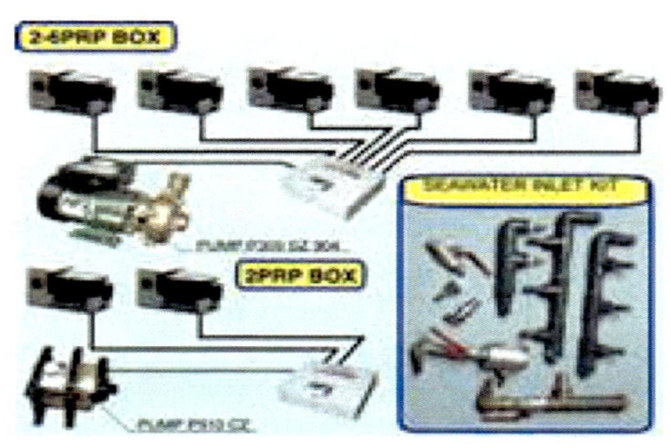

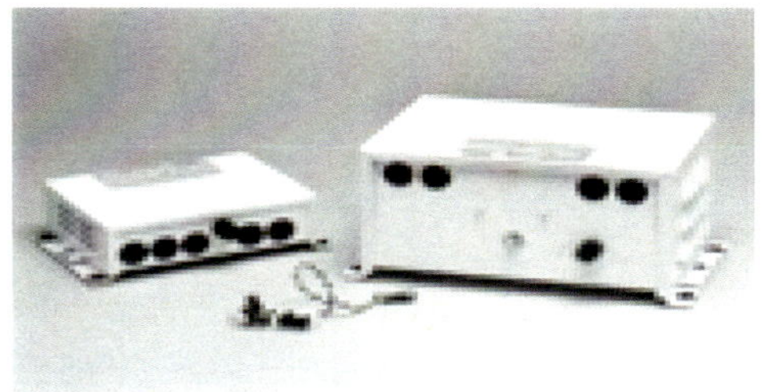

Pump Relay

The sea water discharge from each unit can be individual by each unit or tied together into one or more discharge through hull fittings

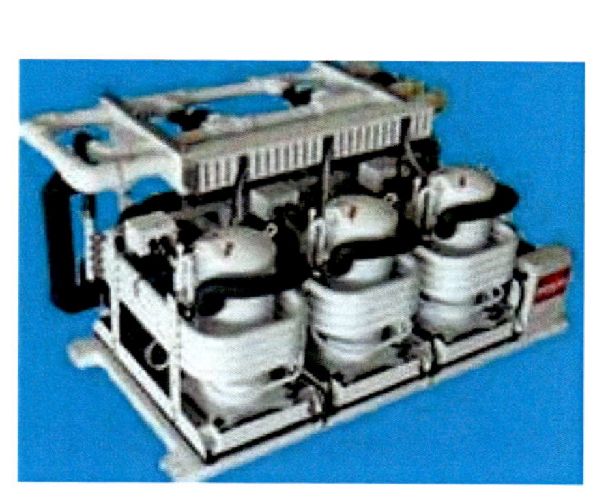

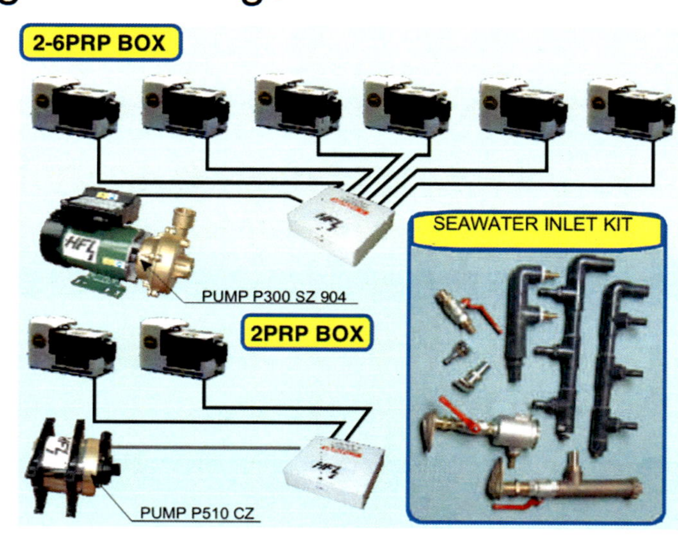

Pump Relay

- In multiple unit installations, I often suggest using one large pump with a pump relay box whereby each unit has a 12/24 volt wire returning to the pump relay box (usually mounted by the pump), and if any one of the units engage, the pump will supply water to all of the units.

- The sea water discharge from each unit can be individual by each unit or tied together into one or more discharge through hull fittings.

Freon

- NEVER CHECK THE FREON PRESSURE! There are no hoses, seals or gaskets to leak or "migrate" Freon molecules into the environment - this is a totally closed system that is of a "critical charge" type, using capillary tubes - even the small amount of Freon lost by connecting and disconnecting the gauges may affect the units performance.

- There is no need to ever check this pressure for if it were to be low on charge there would have to be a leak and a need for the unit to be repaired on the bench - this is an almost non-existent situation.

REFRIGERANT & AIR CONDITIONING

- We have all heard the myth: If your A/C is not working well, all you need is a little more freon
- Adding refrigerant does not necessarily mean your air conditioning is going to work better
- The unit must have a balanced charge. Only a certified air conditioning technician should add refrigerant because the right balance is a must

REFRIGERANT & AIR CONDITIONING

- Refrigerant does not burn up. It does not need to be replaced unless there is a leak in because it is in a closed loop system
- If your unit does have a refrigerant leak however, it is easy to diagnose. Leaking refrigerant is a hazard to people and the environment
- Because the air conditioning element is a closed loop, refrigerant should only be added in the case of a leak

Refrigerant Coils Freeze

- The A C unit will cool for a few hours then stop. If the refrigerant coils freeze inside the air handling unit this is a sure sign you need air conditioning repair
- Ice buildup blocks airflow over the coil
- Low refrigerant causes the unit to operate below freezingz and draw moisture out of the air, forming ice. Freezing ice is not like freezing air

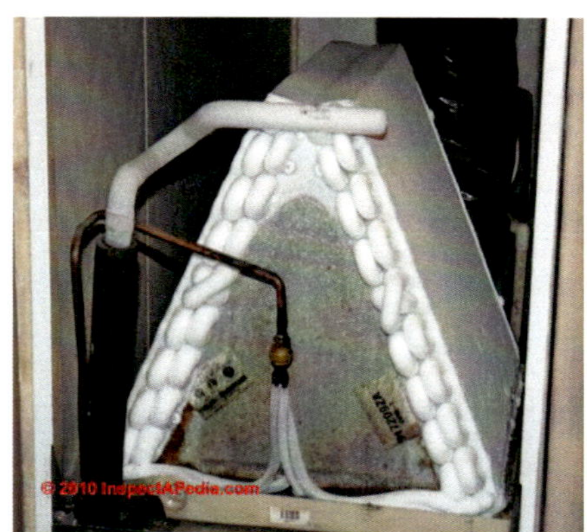

Low Charge of Refrigerant

- A low charge also hurts the efficiency of the air conditioning unit.
- This causes A/C compressors to run above their recommended range which leads to eventual replacement.
- Freon is also linked to the lubrication in some compressors, which can cause the compressor to seize from lack of lubrication

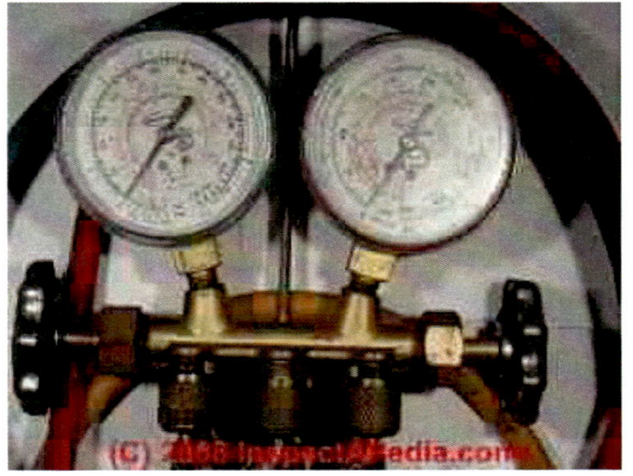

R – 12 or Freon 12

- **Dichlorodifluoromethane** (**R-12**), usually sold under the brand name **Freon-12**, is a chlorofluorocarbon halomethane (CFC), used as a refrigerant and aerosol spray propellant.

- Complying with the Montreal Protocol, its manufacture was banned in the United States along with many other countries in 1994 due to concerns about damage to the ozone layer.

- R-12 was used in most refrigeration and marine air conditioning applications prior to 1994 before being replaced by R-134a, which has a lower ozone depletion potential

Freon

- Refrigerators from the late 1800s until 1929 used the toxic gases, ammonia (NH_3), methyl chloride (CH_3Cl), and sulfur dioxide (SO_2), as refrigerants. Several fatal accidents occurred in the 1920s because of methyl chloride leakage from refrigerators.

- People started leaving their refrigerators in their backyards

- **Freon** is a **chlorofluorocarbon** (**CFC**) is an organic compound that contains Carbon, Chlorine and fluorine.

Freon

- The CFCs are a group of aliphatic organic compounds containing the elements carbon and fluorine, and, in many cases, other halogens (especially chlorine) and hydrogen. Freons are colorless, odorless, nonflammable, noncorrosive gases or liquids.
- In 1930, Thomas Midgley held a demonstration of the physical properties of Freon for the American Chemical Society by inhaling a lung-full of the new wonder gas and breathing it out onto a candle flame, which was extinguished, thus showing the gas's non-toxicity and non-flammable properties.

Freon

- Only decades later did people realize that such chlorofluorocarbons endangered the ozone layer of the entire planet
- CFCs, or Freon, are now infamous for greatly adding to the depletion of the earth's ozone shield. Leaded gasoline is also a major pollutant, and Thomas Midgley secretly suffered from lead poisoning because of his invention, a fact he kept hidden from the public.

Freon – R-22 and Freon 12

- This colorless gas is better known as **HCFC-22**, **R-22**. It was once commonly used as a propellant and in air conditioning applications. These applications are being phased out due to ozone depletion potential and status as a potent greenhouse gas.
- In later years R-22 was replaced by Freon-12.
- Freon-12 was very widely employed as a refrigerant and also a propellant in spray cans until 1994, when it was banned under the Montreal Protocol as an ozone-depleting chemical

Marine Refrigerants

- In summary, the 4 commonly used refrigerants that you can find today are:
  - CFCs : were developed more than 70 years ago and are harmful to our respiratory systems and the ozone layer
  - HCFCs : Their production are scheduled to be phased out totally in 2030 (R-22, R-123)
  - HFCs : R-134a and R-124 are used in Industrial systems that are specially designed for its use
  - Refrigerant blends : R-410A ,R-407C are used as R-22 replacement
- The future commonly used refrigerants will be in the last two categories. Among the currently widely used ones are R-134a, R407C and R410A

Where can you buy freon to charge your air conditioner?

- R12 is almost impossible to buy, unless you go to a large-scale boat parts supplier. But they won't sell it to you unless you are a certified A/C technician.

- You can buy R134a easily at other marine stores, but be careful! If your boat's A/C system is depleted of all refrigerant, you have a serious leak that must be fixed. Also, a vacuum pump must be used on the system (to remove air) before you introduce R134a.

- Depends on what type of freon your boat uses. If it is pre 1994ish you will probably need R-12 freon

EPA Certification

- The United States Environmental Protection Agency (EPA) has required that all persons handling refrigerants should be certified. If you are not certified you can't legally buy refrigerants.

- Three types of certification are available. The Section 608 Type I, II and III .

- Type I Certification – Can only work on Small Appliance (5lbs or less of refrigerant)

  Type II Certification – Can only work on Medium, High and Very-High Pressure Appliances.

  Type III Certification – Can only work on Low-Pressure Appliances.

Free EPA Certification Software Download

- You can take the test right from your home by downloading Mainstream's new testing software. This software is intended to provide you with new study tools and the means to take your A/C certification tests right from home without the risk of losing your internet connection.

- Simply click on the download link below, study the material using the study guides and practice exams included with the software, take the test, and come back to www.epatest.com to grade an exam.

Future Refrigerants

- Effective January 1, 2010, manufacturers of air-conditioning and refrigeration systems were required to stop using the traditional "R" series refrigerants like R-12 and R-22.

- In the marine realm, these have been replaced with a new "R", 410A which meets all of the EPA requirements.

- The older R-12 and R-22 will gradually be depleted between now and 2020 when they will become essentially banned substances

Freeze - 12

Freeze-12 is one of the only (*EPA allowed*) direct replacements for R-12 or R-134a. Performance and efficiency is very similar to that of R-12 (Freon) and better than that of 134a

REFRIGERANT PHASE-OUT

- Many of the current forms of refrigerants used today are being phased out based on concern for depletion of the ozone layer

- Portables use R-22, which has been deemed acceptable for use by the EPA until the year 2010. By that time, an ozone-friendly refrigerant that can be easily substituted for R-22 will be readily available

REFRIGERANT & AIR CONDITIONING

- Some Marine/Residential units use a R-22 coolant. Prices are rising as R-22 is fazed out because of its ozone depleting properties.
- Phase out begins this year and is to be completed by 2020
- The Marine tech should find the leak when recharging an A/C system. A certified A/C system should not need to be refilled more than once a year.

Future Refrigerants

- If you own a boat with an older system installed, you've basically got another 9 years before it becomes totally obsolete

- It's not just different refrigerant that will make any system built prior to January 1 2010 obsolete

- Once an older system loses its refrigerant charge, a service technician will not be able to simply repair the leak and recharge the system. Why? Because the new systems utilizing the R-410A refrigerant are running at much higher pressures than the old systems

Aluminum vs. Copper: the great condensing coil debate

- Copper tubes with aluminum fins, which are the usual choice for condensing coils, are superior to aluminum because of copper's superior strength, reliability, ease of maintenance, and excellent heat transfer characteristics.

- Some manufacturers continue to use aluminum for their condensing coils.

- Two of the biggest problems are that aluminum coils are more difficult to maintain and almost impossible to repair

Self-Contained Marine A/C

Self-Contained Marine A/C

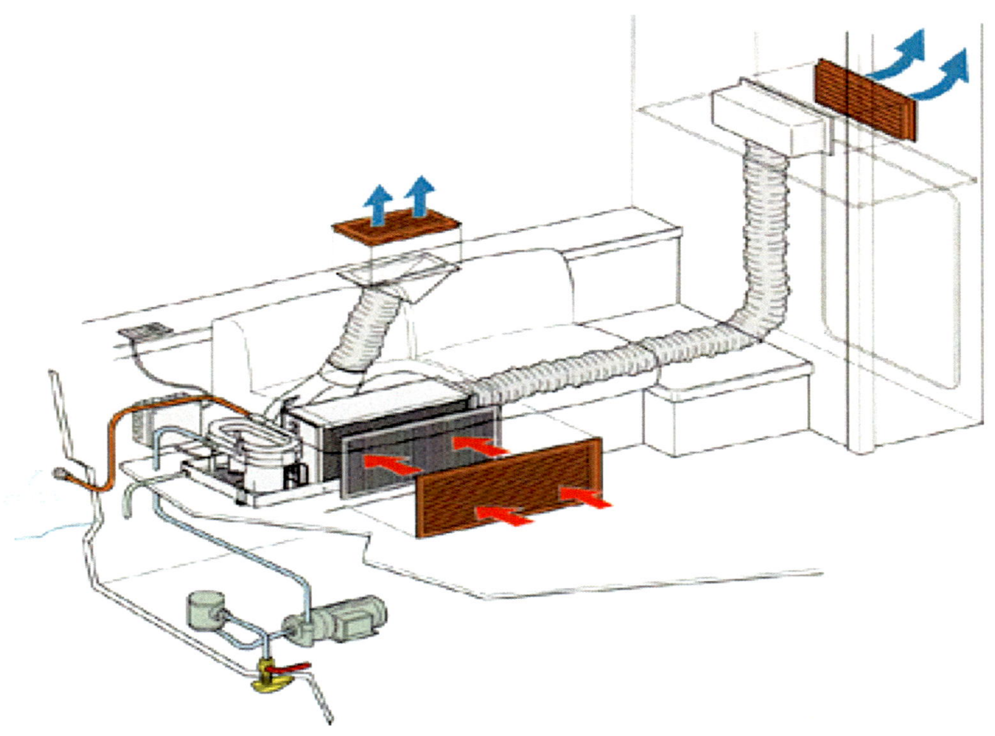

Leak Detectors

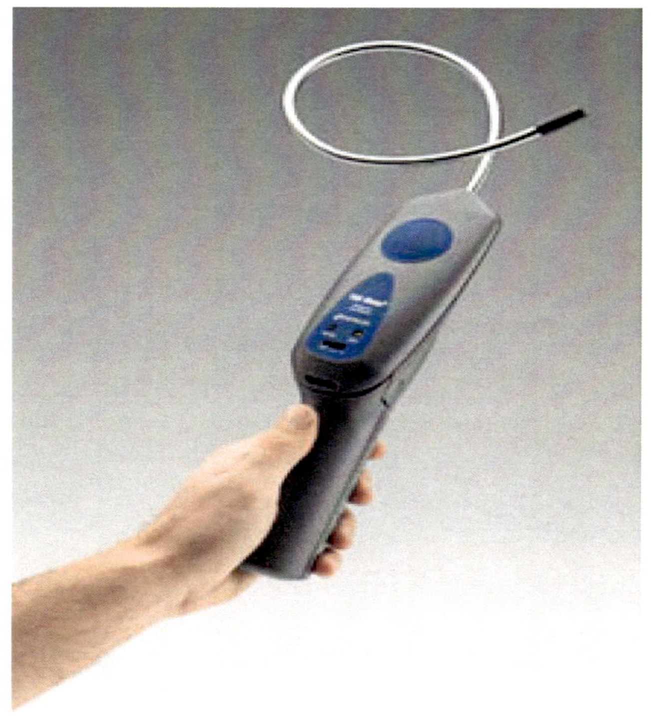

Leak Detectors

- There are just two types of electronic leak detectors. There are those that use a heated diode, and those that rely on corona discharge

- On the corona discharge type, anything that breaks that corona barrier will set off the detector. You name it - dust, moisture, refrigerant (sometimes), solvents, etc.

- The heated diode type detectors are less prone to false alarms from moisture and shop chemicals. More importantly, heated diode leak detectors are also much more sensitive to R-134a.

Leak Detectors

- Today, there are many heated sensor units to choose from, and most are very economical to own.

- When researching which detector to buy, consider sensor life expectancy, and unit sensitivity during comparison. These specifications should be mentioned in sales literature.

- **Other factors to consider when choosing a leak detector:**
  - sensor replacement cost
  - warm-up time
  - power supply

Refrigerant recovery process

- **Refrigerant recovery takes place mainly at three stages of air-conditioning equipment:**
- During servicing when partial or complete refrigerant charge is removed from the system.
- When the system is converted to run with a new refrigerant.
- When the equipment is disposed

Refrigerant Recovery Machine

The refrigerant recovery machine has come a long way in a few short years. On the surface, it is a machine that simply takes refrigerant out of a system and puts into a tank

Vacuum Pump

- After the refrigerant has been removed, and before the a/c system is ready to receive the new refrigerant, use a vacuum pump to remove moisture and air from an a/c system

- Modern systems are built tighter so charges are more critical. That means these systems have a greater sensitivity to moisture and other contaminants, making thorough evacuation very important

Moisture

- Moisture in a refrigerant system causes the most problems. Moisture can cause freeze-up, create acid that increase the corrosion of metals and reduce the lubricating ability of the oil.

- The latest recovery systems integrate a recovery machine with a vacuum pump .

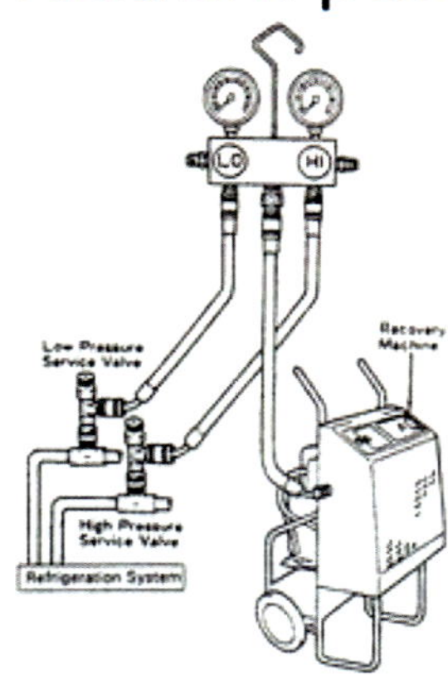

Recovery Tank

One of the simplest and most cost-effective ways to achieve this is to use a 30- or 50-pound tank in-line on the system's input line. From the system, connect to the liquid port of the tank. Then, from the vapor port of the tank, connect to the input of your recovery machine

Recovery Process

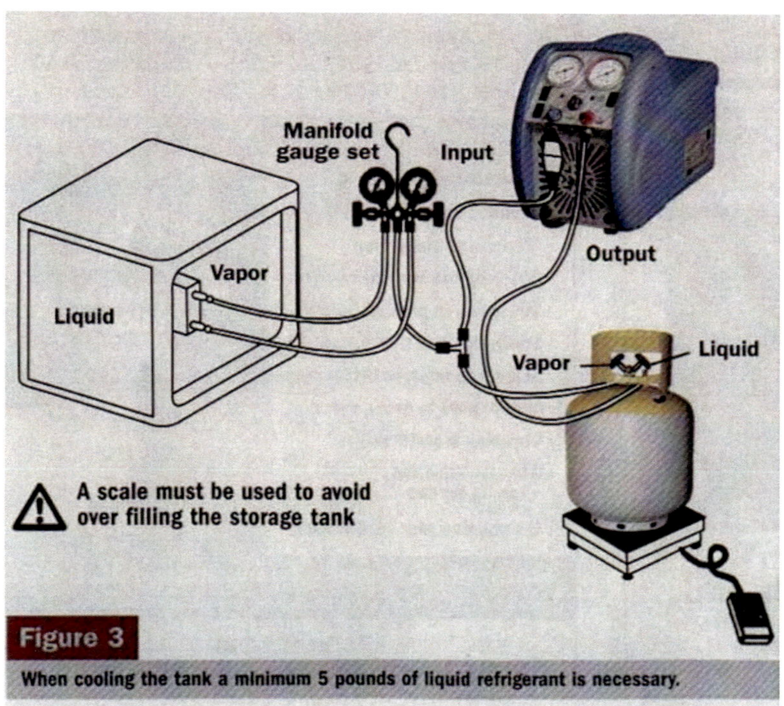

Recovery process

- Refrigerant recovery systems all work using a vacuum or compression pump

- A vacuum pump pulls the refrigerant out of the donor's compressor and into the recovery tank

- Current EPA standards require that 95 percent or more of the old refrigerant be evacuated from the donor. Recovery machines contain a "float cable" that signals the pump to shut off as the recovery tank becomes full

Recovery process

- Some recovery machines actually contain compressors rather than simple vacuum pumps that essentially enlarge the compression cycle to draw the refrigerant out into the recovery tank.

- Hoses used to drain systems are either 1/4 or 1/2 inch in diameter, depending on the specific refrigerant used. Technicians must match recovery equipment to the equipment to be serviced

Speeding up the recovery process!

- Use the shortest hoses possible for the job. Long hoses increase the recovery time

- Always identify the refrigerant you are recovering. This will minimize cross contamination and help you plan for the amount of refrigerant you will be recovering

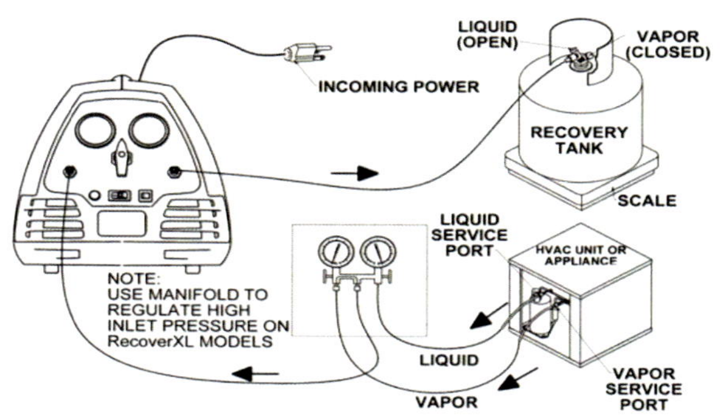

Speeding up the recovery process!

If possible, recover from both the high and low side service port on the system from which you are recovering

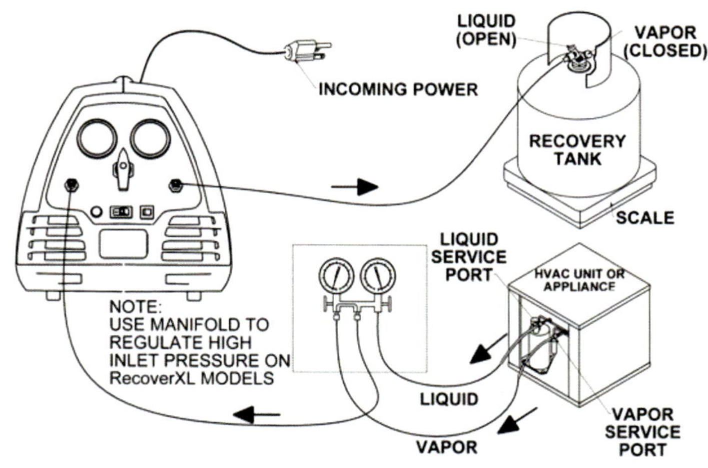

Refrigerant Piping

Prepare pipe and fittings with care, particularly when cutting copper tubing or pipe to prevent filings or cuttings from entering the pipe

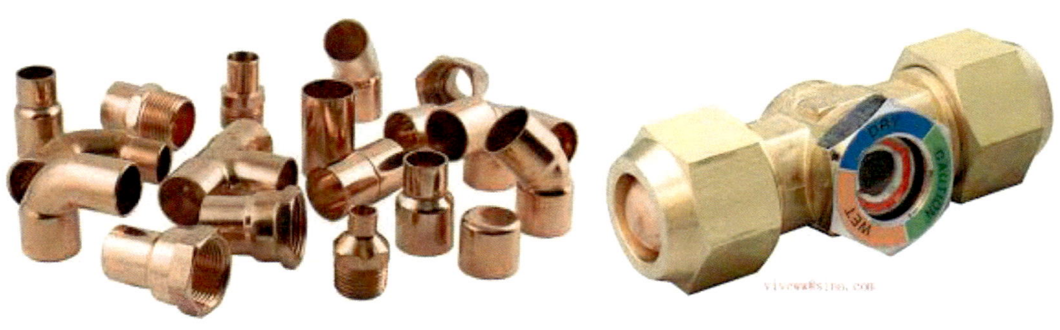

Refrigerant Piping

- The tube should be cut square, and all burrs and dents should be removed to prevent internal restrictions and to permit proper fit with the companion fittings.

- The use of a hacksaw should be avoided whenever possible. When making silver-solder joints, brighten up the ends of the tubing or pipe with a wire brush or crocus cloth to make a good bond.

TK-404

Soldering Copper Pipes

Acid should never be used for soldering, nor should flux be used if its residue forms an acid

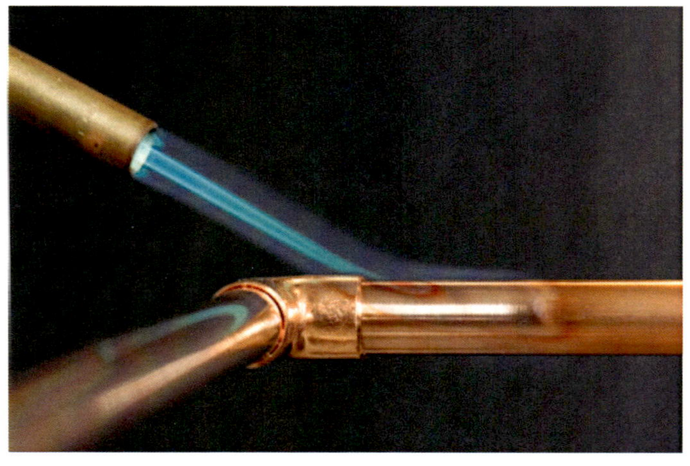

Soldering Copper Pipes

- Use flux sparingly so no residue will enter inside the system and eventually be washed back to the compressor crankcase.

- If tubing and fittings are improperly fitted because of distortion, too much flux, solder, and brazing material may enter the system

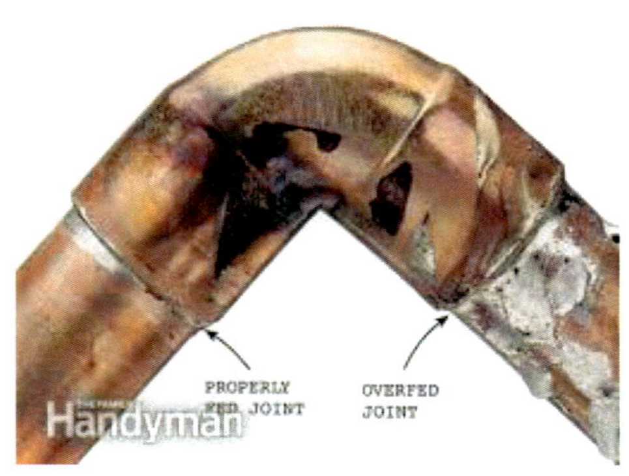

Refrigerant Piping

- The temperature required to solder or braze pipe joints causes oxidation within the tubing. The oxidation eventually will be removed by the refrigerant flow after the system is in operation.

- The oxide breaks up into a fine powder to contaminate the lubricant in the compressor and to plug strainers and driers.

- All joints should be silver-soldered and kept to a minimum to reduce leaks.

Flare Joints

- SAE flare joints are generally not desired, however in marine installations flare joints is in some cases the only way to solve a leak issue, care should be taken in making the joint.
- The flare must be of uniform thickness and should present a smooth, accurate surface, free from tool marks, splits, or scratches.

Flare Joints

- The tubing must be cut square, provided with a full flare, and any burrs and saw filings removed
- The flare seat of the fitting connector must be free from dents or scratches

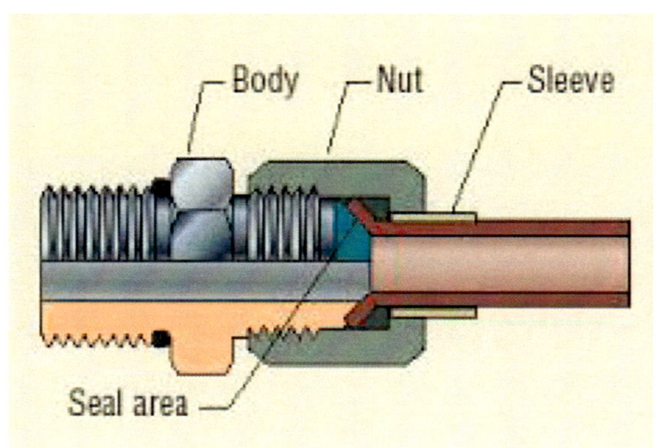

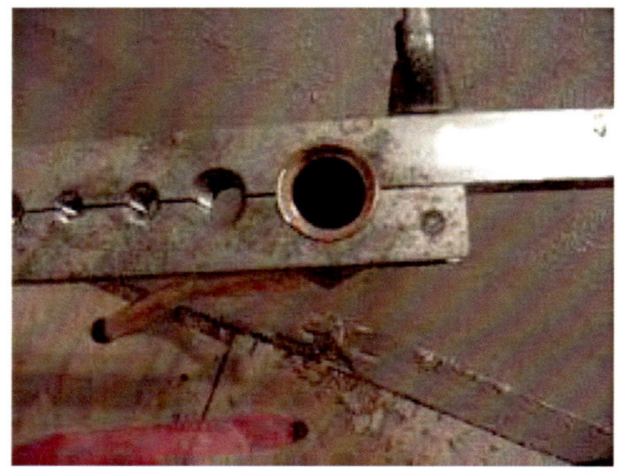

Flare Joints

The flare can best be made with a special swivel head flaring tool, available as a general stores item, which remains stationary and does not tear or scar the face of the flare in the tubing

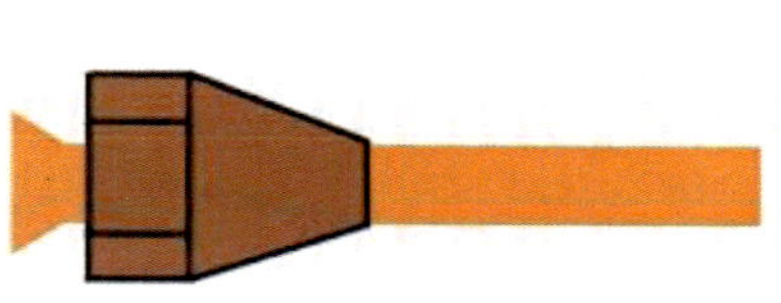 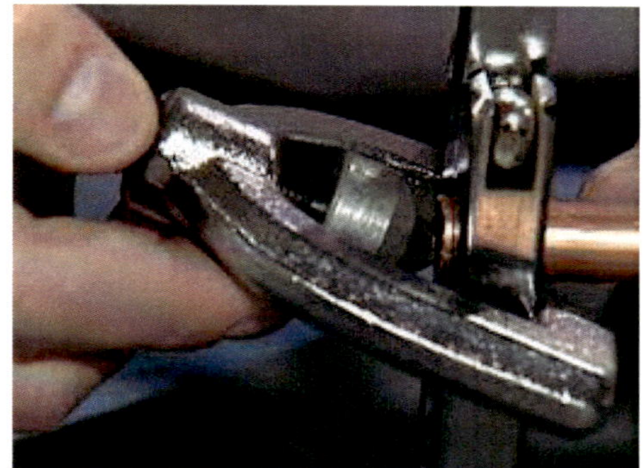

Flare Joints

- Oil should not be used on the face of the flare, either in making up the flare or in securing it to the fitting, since the oil will eventually be dissolved by the refrigerant in the system and cause a leak through the displacement of the oil

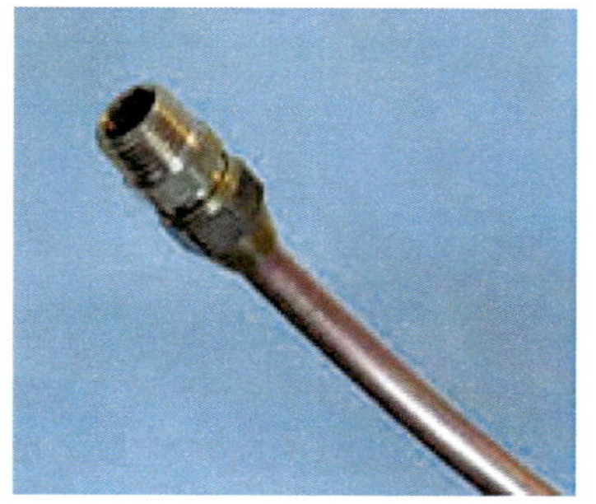

Bending Pipes

- Where pipe or tubing has to be bent, bends should be made with special tools designed for this type of work. Do not use rosin, sand, or any other filler inside the tubing to make a bend.

- Threaded joints should be coated with a special refrigerant pipe dope. In an emergency, use a thread compound for making up a joint; remember R-12 and R-22 are hydrocarbons, which dissolve any compound containing oil.

- Thread compounds should be applied to the male part of the thread after it has entered the female coupling one and one-half to two threads to prevent any excess compound from entering the system.

Electrical Installation

- Never use solid core wire on any vessel because the motion of the vessel may eventually cause a failure and can be a serious fire hazard.

- Be sure to use the proper size wire for our equipment. Undersized wire or poor dock/ship connections will result in a voltage drop that may damage the equipment and be a fire hazard.

- Use a minimum 12 gauge power supply wire - never use 14 gauge for even our smallest units if wired 115 volt.

Electrical Installation

- Be sure the breaker capacity is less than the current capability of the wire.

- Many marinas have inadequate wiring -we suggest you test the dock line voltage during peak loads -it may be 120 volts on a cool weekday morning and 90 volts on a hot weekend afternoon.

- With 120V systems, the black wire is your line (L), which goes to the circuit breaker, the white wire is your neutral (N), which goes to the neutral buss bar, and the green wire is always ground (G), which goes to your ground buss bar.

Air Conditioning Capacitors

This section is measured in micro-farads, voltage rating and listed as either RUN CAP, DUAL CAP, START CAP or HARD START CAP

Air Conditioning Capacitors

The Dual Round and Dual Oval Capacitors perform the same

Condenser / Capacitor

- If your air conditioner has stopped running, a possible problem is the failure of the starting capacitor found on the outside compressor/condenser unit.
- If that unit has electrical power but the compressor and/or its cooling fan are not running, one of the components to check (and that is easy to replace) is the starter.

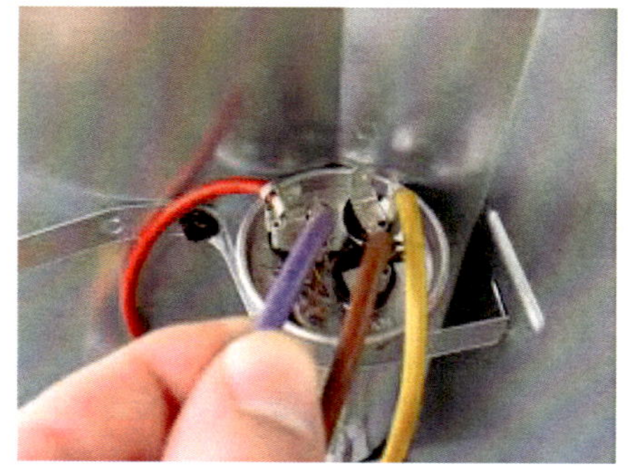

Start Capacitor

- **A/C Compressor starts, runs just briefly, then shuts off**: this condition too may be solved by adding a starting capacitor, but watch out: it may also be a sign of a compressor near end of life

- When an electrical motor is having trouble starting, such as an air conditioning compressor motor ,blower motor, a refrigerator motor or a freezer motor, or even a fan motor, the repair technician may install a simple and inexpensive starting capacitor

Start Capacitor

The starting capacitor is a simple electrical device which can give an extra voltage jolt or "boost" to get the hard-starting motor spinning

Start Capacitor

- The starting capacitor is oval in cross section, but most replacement and many newer air conditioner motor starting capacitors are simply cylindrical in shape

- Look for starter capacitors in your air conditioning equipment in the engine room condenser where a starter capacitor may be used to aid compressor motor start-up

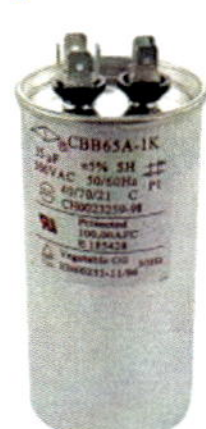

Starting Capacitors

- Its function is to assist in developing high torque to get a motor turning over. It increases starting torque by aprox. 250%.

- When replacing capacitors, one of equal value or more expressed in farads should be used.

- Never use replacement with a lesser value, doing so may overheat the capacitor and / damage the motor

- In general starting capacitor is oval in cross section

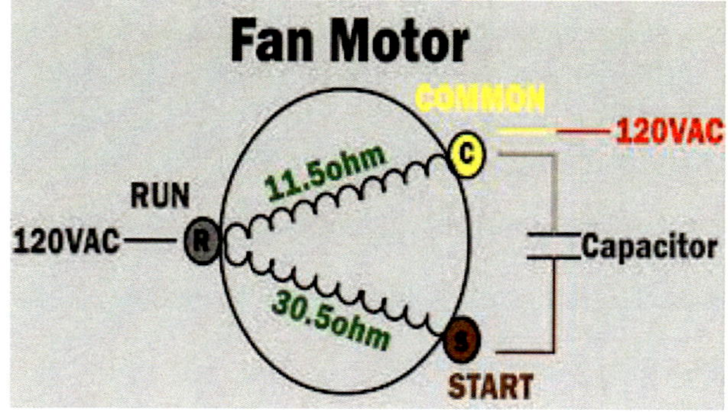

How a Starting Capacitor work

- The starting capacitor, by "accumulating" a large electrical charge inside the capacitor. Traditionally capacitors were also called "accumulators" for this reason.
- The capacitor's electrical charge is released at motor start-up time, gives the compressor motor or other electrical motor a boost for starting.
- This "boost" can be particularly needed if an air conditioner is suddenly switched off and back on when it has been operating

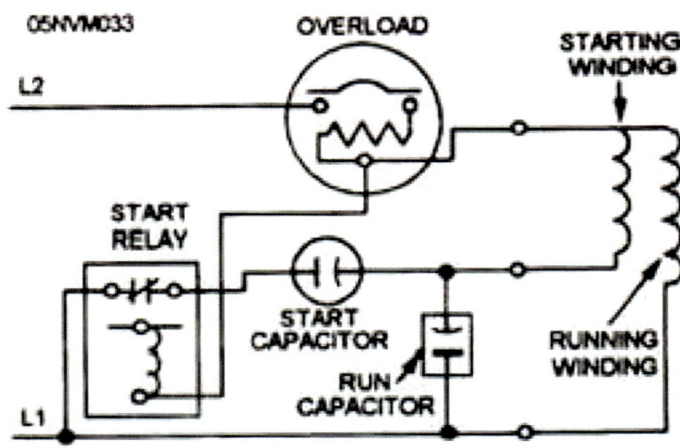

Where to Find or Locate the Starter Capacitor

Look for starter capacitors in your air conditioning equipment in the outside air handler where a starter capacitor may be used to aid compressor motor start-up, and look for a smaller motor starting capacitor on fan motors, both in the outside compressor/condenser and indoors in the air handler unit's blower compartment

Causes of Hard-Starting Air Conditioners, Refrigerators, Freezers

- **Low line voltage** supplied to the equipment or excessive power drop on a long circuit wire

- **Un-balanced cooling or refrigeration systems** - such as a compressor motor that is having trouble starting when the refrigerant pressures are high on one side and low on the other side of the cooling system. This hard starting condition happens when a compressor is turned off in the middle of an "on" run cycle

Causes of Hard-Starting Air Conditioners, Refrigerators, Freezers

- **Old, aging compressor motors** or other electrical motors that are at or near the end of their life may have trouble starting and may be able to function for some additional time given the "help" provided by a starting capacitor
- **Troubles with an electric motor** such as used on an air conditioner air handler, condensing coil fan, or a well water pump

Causes of Hard-Starting Air Conditioners, Refrigerators, Freezers

Bad or failed starting capacitor: the air conditioner compressor (and some other electric motors) may already have a starting capacitor installed, but the starting capacitor might have failed, causing the air conditioner compressor to start with difficulty or not at all. <u>A bad starter capacitor can also disable the fan in the outdoor compressor/condenser, or the blower fan in the indoor air handler unit</u>

How to Use a Multimeter to Test a Motor Starting Capacitor

- Use a VOM in ohms setting to check resistance across the capacitor.
- If the meter does not move (no current flows) the capacitor is "open". If there is zero resistance the capacitor is shorted.
- A service technician can test for a failing motor by measuring the current draw in amps during start and run, and by comparing the result to information on the motor's data tag

Once the capacitor has been discharged

- Charge the capacitor by using the sense current the meter puts out when set to ohms. You should observe a rapidly rising resistance before the meter indicates over range/infinity.
- Disconnect the test leads, and switch over to volts. Then, reconnect the test leads. A voltage reading should be observed, approaching zero.
- If the capacitor doesn't hold a charge, or the resistance reading never approaches infinity, it probably needs replacement

Checking Blower, Fan Motor or Outdoor Compressor Fan Motor

- Unplug electrical connectors at the fan motor. Measure the resistance between each lead wire with a multimeter. The multimeter should be set in the X1 range. For accuracy, don't measure when the fan motor is hot, allow it to cool off.
- When the resistance between each lead wire are those listed in the specifications for your equipment the fan <u>motor should be normal</u>.
- Zero resistance or infinite resistance are indicators of a problem

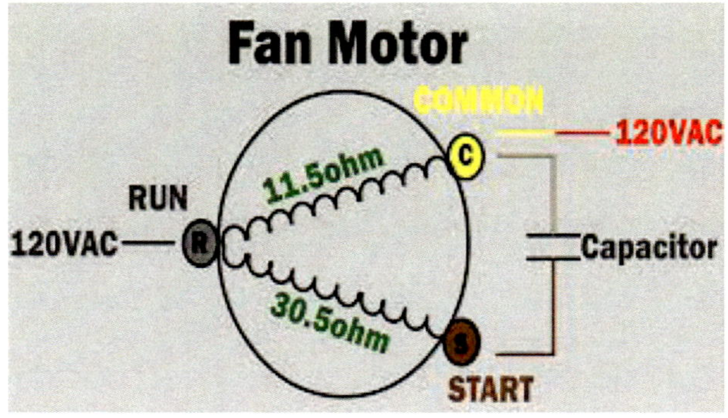

312

How to Install a Starting Capacitor

- **Remove the power supply cord** from the electrical outlet.
- **Remove the old starting relay**, leaving the old overload protection in place.
- **Push the wire with the one single pin terminal onto the "start" terminal** of the air conditioning compressor.
- **Push the other wire with the pin terminal onto the "run" terminal** of the air conditioning compressor

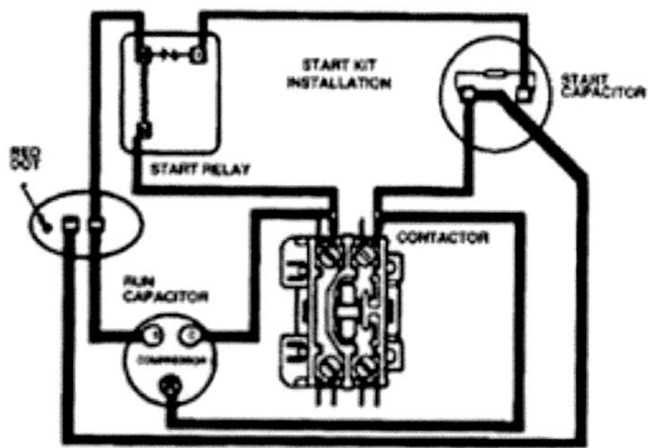

How to Install a Starting Capacitor

- Connect the line from the old starting relay to the spade terminal on the "run" wire .
- Restore electrical power
- **When testing a compressor, one must discharge the capacitor first!** It'll otherwise have enough power stored on it to be at least very painful
- Some systems will automatically discharge the capacitor, but shorting its leads with a screwdriver (after verifying that the power's off) is a safe way to ensure that you won't get shocked. Motor starting capacitors can hold a charge for days

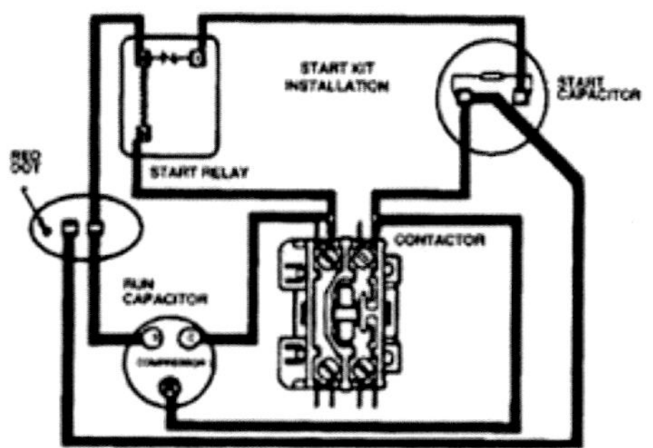

Thermostat

- There are two types of thermostats : Programmable Digital and Mechanical.
- Each day the digitals are widely recommended because they are more accurate

 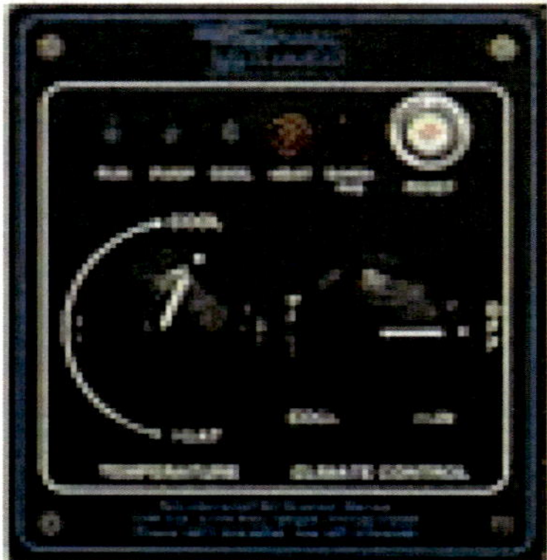

Thermostat and Temp. Sensor

The thermostat sensor is located at the suction side of the evaporator. The other end is plugged in the thermostat control box

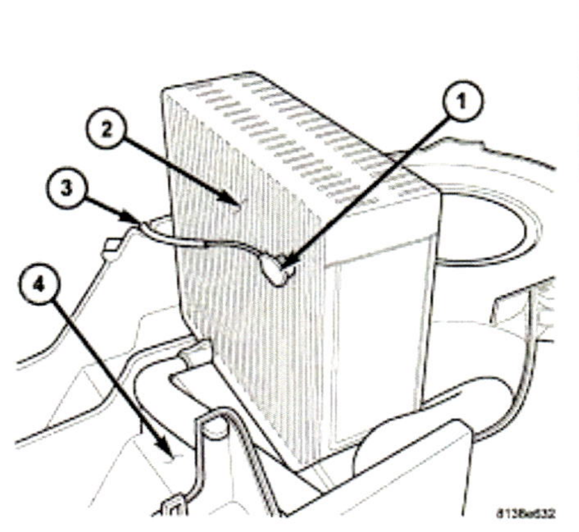

Mechanical Thermostat

The sensitive Bulb of the mechanical thermostat is located in the same place at the evaporator coil

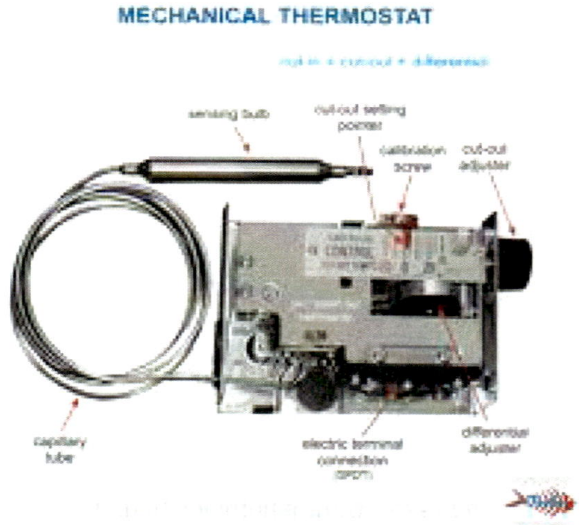

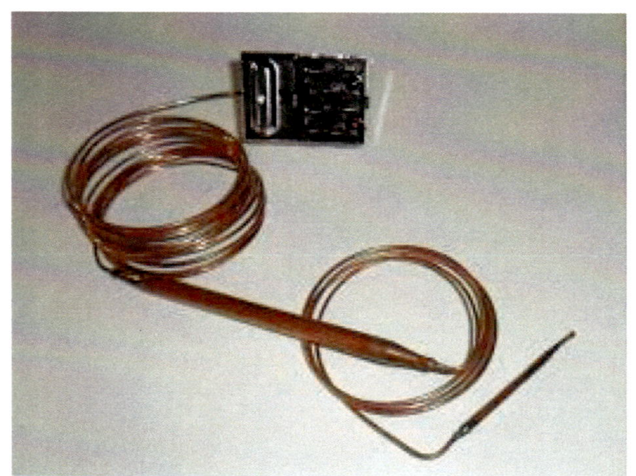

Thermostat troubleshooting

Raise the thermostat heat settings slowly. The thermostat should emit a clicking sound and the furnace should also make a sound. If the furnace does not make a sound, the thermostat is not sending a signal and needs to be replaced

Thermostat Troubleshooting

Turn off the breaker to your furnace. Remove the cover from the thermostat to expose the wiring. Because the thermostat operates on low voltage, there is no risk of electrocution

Thermostat Troubleshooting

Take a picture with your phone or digital camera of the wiring inside the thermostat to show how they connect. Loosen the screw label "R" or "Rh" and the screw labeled "W." Remove the wires from beneath the screw

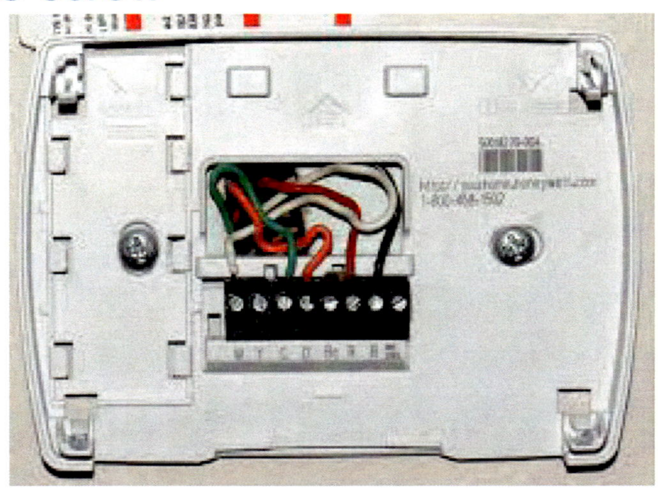

Thermostat Troubleshooting

Twist the ends of the wires together. Turn on the evaporator breaker. If the evaporator comes on, the thermostat is not sending a signal and needs replacing. If the furnace does not turn on, the furnace needs repair

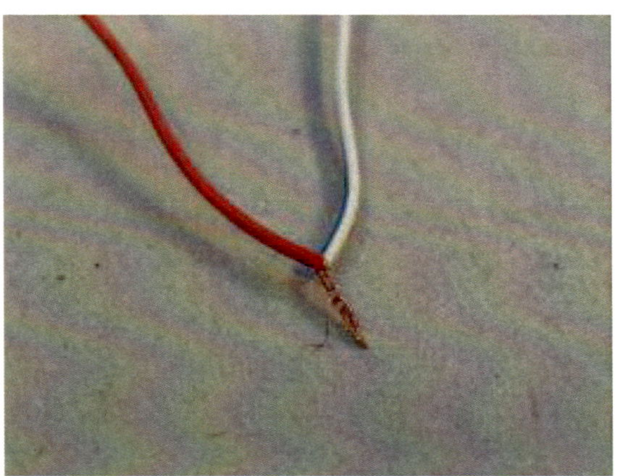

Duct Material and Routing

In marine applications flexible ductwork, both insulated and un-insulated is the most common type used. Whether it's insulated or not depends primarily on where it's routed

Duct Material and Routing

- Ductwork routed through hot areas of the boat should be insulated to minimize heat transfer
- The rule of thumb is to keep ducts as short as possible, as straight as possible and as large a diameter as possible

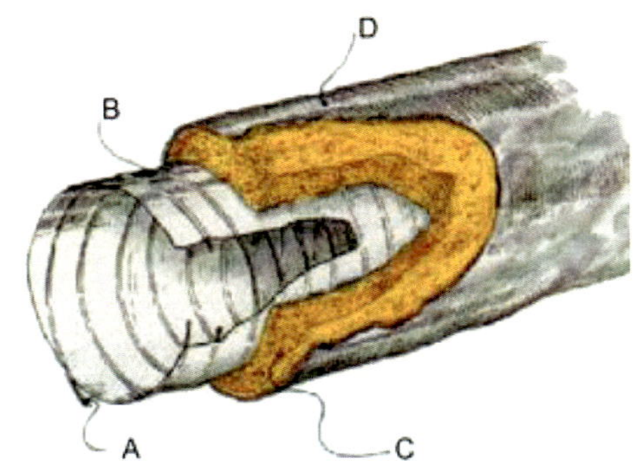

Duct sizing & Grills

- Typically the return air duct should be 20% larger in diameter than the cool air supply duct
- It is recommended that a wooden grille be used for the discharge air flow to minimize condensation on the frame

Air Filters

- Dirty air filters can cause system overload and may even manifest as a constantly tripping circuit breakers
- They will cause low suction pressure, which will contribute to the overload and again, may trip a low pressure switch in the system
- Operating without an air filter is a bad idea as well because the net result will be an evaporator assembly that will be really difficult to clean

Water Pumps

- High capacity, centrifugal pumps.
- 250 -10,800 gph capacities.
- 115V and 230V, 60 Hz; 240V, 50 Hz; and 220/440V, 3 ph. models.
- Seawater grade construction, with glass filled polypropylene or bronze pump heads
- Submersible or air cooled motors
- High head pressure models available
- Self-priming pumps available

Air Conditioner Sizing

- Air conditioner size is rated using the number of **BTU's** or British Thermal Units of heat that it can remove per hour.
- **1 BTU** is the amount of energy to raise the temperature of one pound of water by one degree Fahrenheit
- You may also hear ac size referred to in "tons" which is the equivalent of 12,000 BTU's

Conversion for heat and power units

- **1 kW is 0.7457 horsepower.**

- **1 kWhr is 3,413 BTU/hr (British Thermal Units).**

- **1 ton is 12,000 BTU/hr**

Square feet method for air conditioner sizing

- **Quantity Per Sensible heat gain**
  1 ton = 500 ft^2 of floor area
- **Heat gain from person** 380 BTU per person
- **Heat gain from cooking** 1,200 BTU per kitchen
- Increase BTUs by 600 for each additional person (using room) over 2 persons
- Increase BTUs by 10% for sunny areas

Duct sizing & Grills

Capacity (BTU/hr)	Duct (in)	Return Air Grill (Sq.in)	Discharge Grill (Sq.in)
5,000	4"	60	30
7,000	5"	80	45
10,000	6"	100	60
12,000	6"	130	70
16,000	7"	160	80
18,000	7"	200	100

Air Cond Capacity

Unit Capacity	Volt	Duct Diam	Price
6000 BTU	120 V	4"	1,524
10000 BTU	120 V	6"	1,750
12000 BTU	240 V	6"	2,200
16000 BTU	240 V	8"	2,600

Marine DC Air Conditioning

- 5,000 BTU/hr of air conditioning in hot tropical climates
- Direct 12 volt or 24 volt DC input - no inverter required

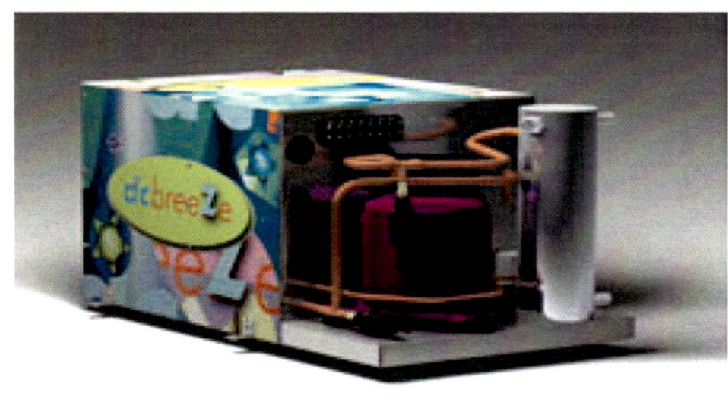

Chapter 11
Watermaker Systems

- Watermaker Components
- Types of Watermaters
- Power Supply
- Sizing a Watermaker
- Flushing the System

Types of Watermaker

- They come in **modular setups**, Self Contained or Portable and handheld units .

- There are watermakers for boats, watermaker for yachts, watermakers for sailboats and for any size boat

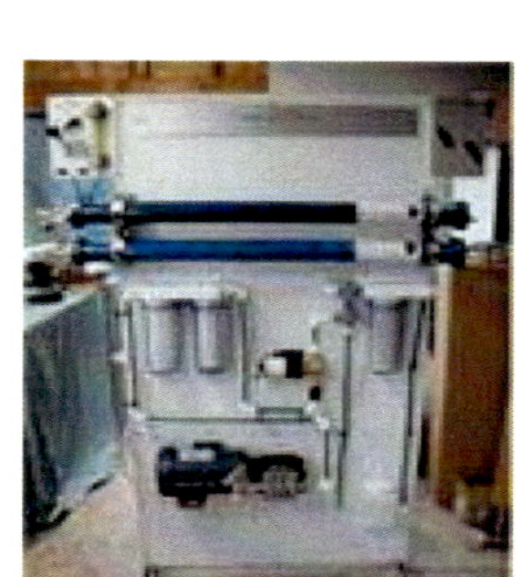

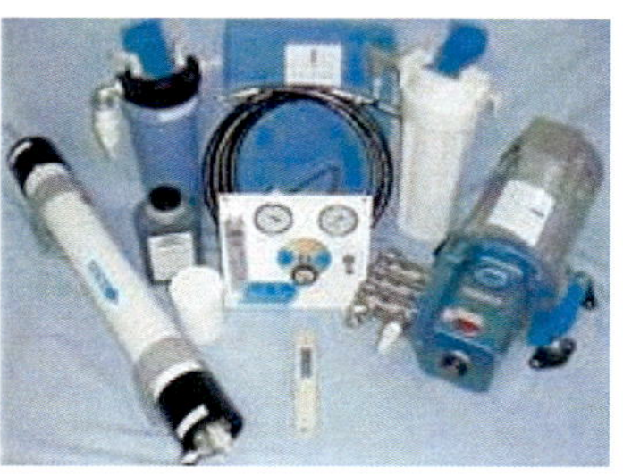

Watermaker Components

- A typical marine watermaker installation involves:
  - A seawater supply ,(Thru-hull, Valve and hoses)
  - The preliminary filtration to remove weeds and large contaminants (Primary and Secondary Filters)
  - A low pressure pump to push the water through the particle filter which remove particles down to 5 microns
  - A high pressure pump to supply the RO membrane. Fresh water from the membrane is then sent to the water tank

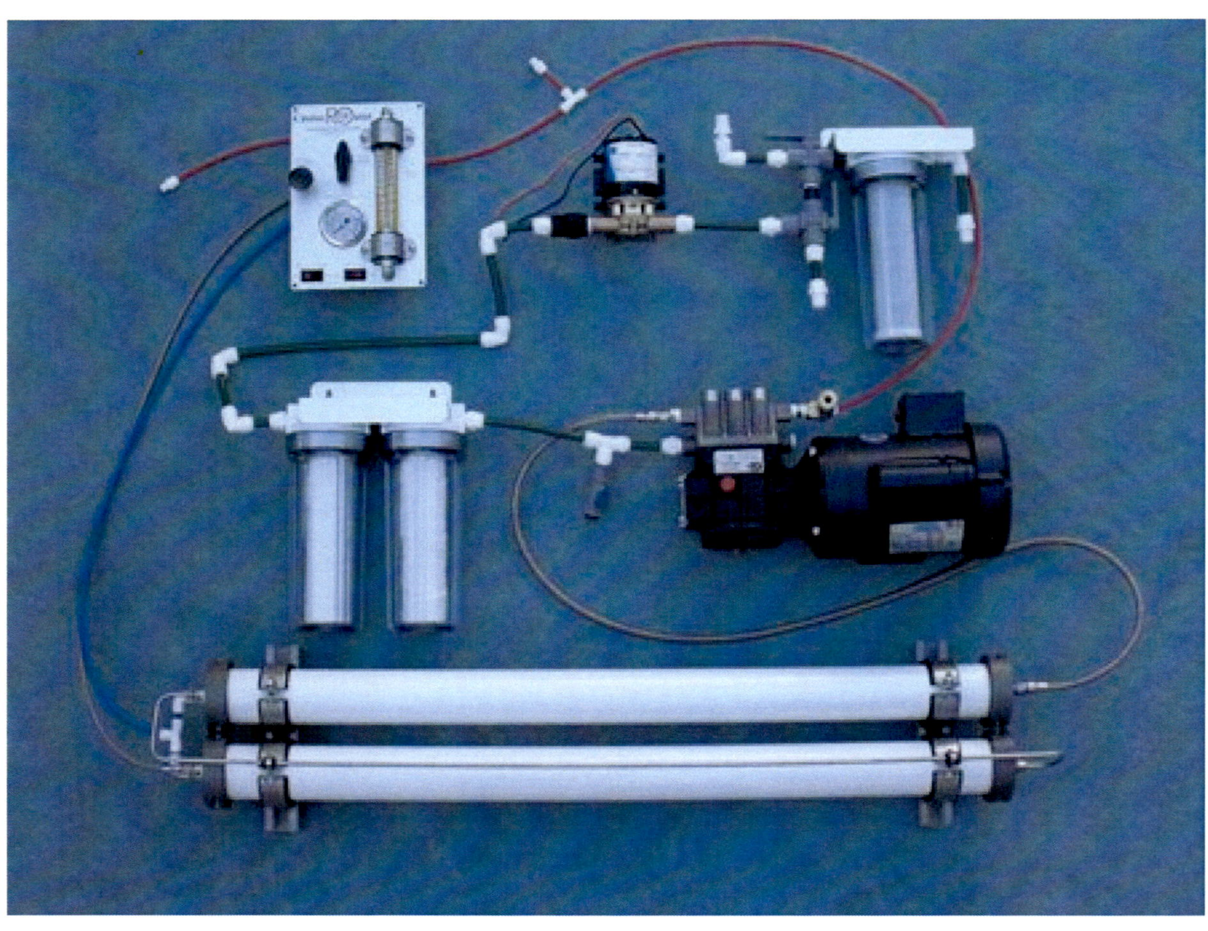

Control Panel

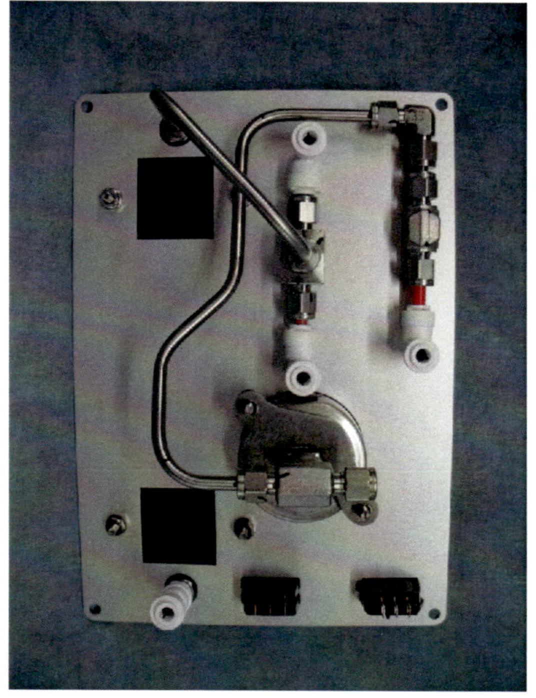

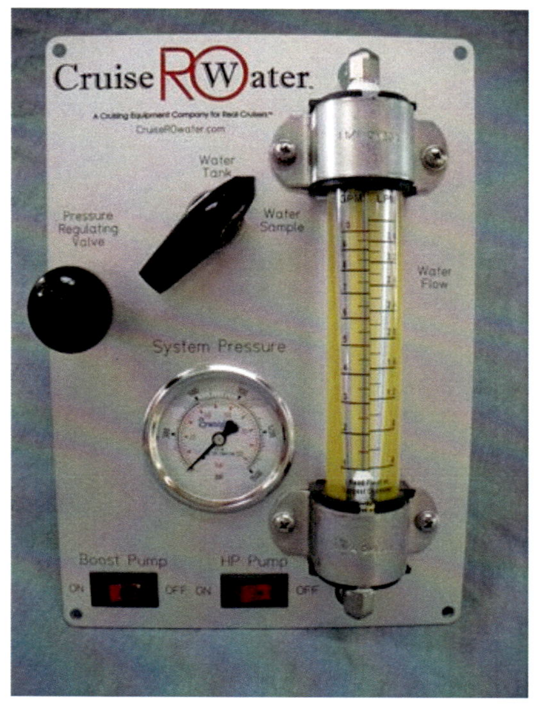

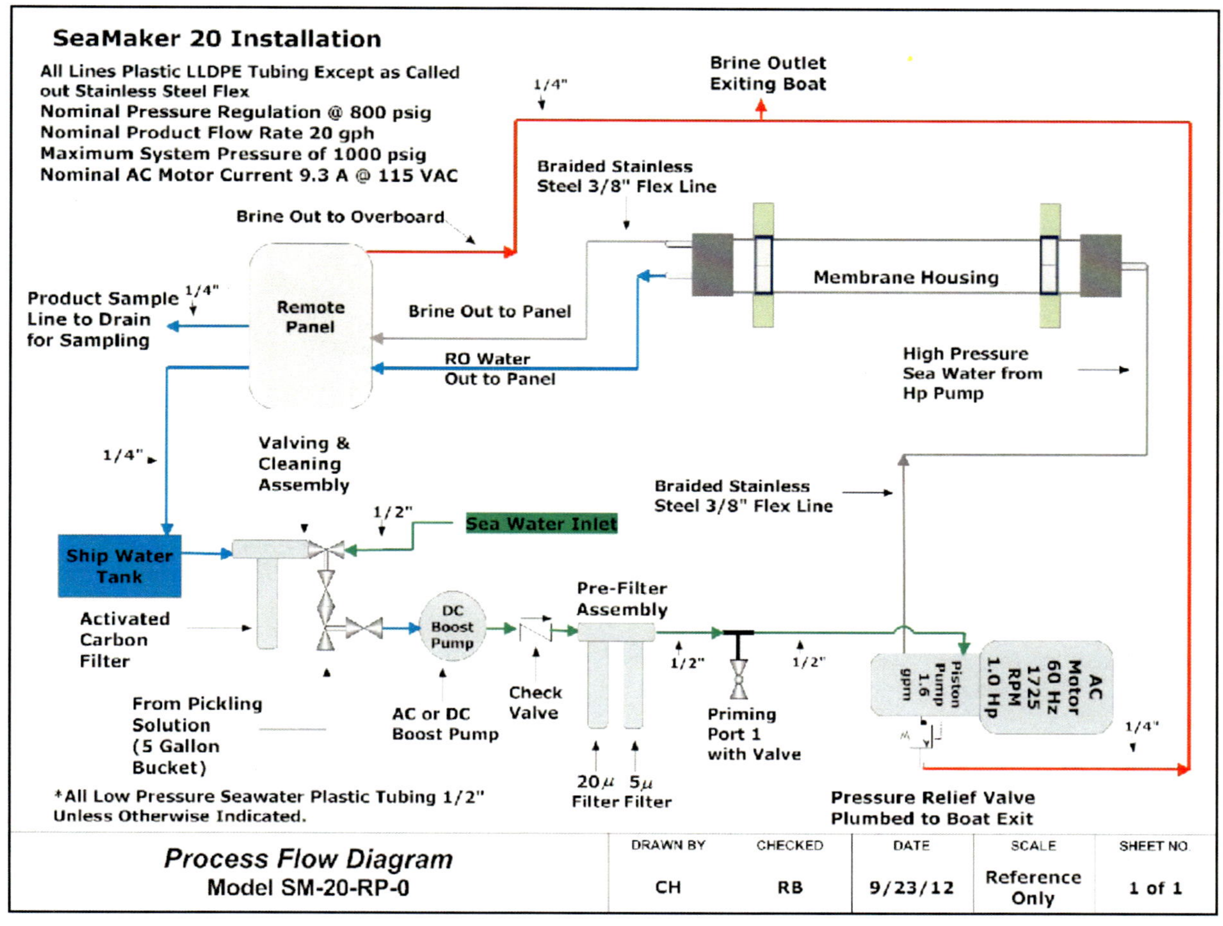

Process Flow Diagram Model SM-20-RP-0	DRAWN BY	CHECKED	DATE	SCALE	SHEET NO
	CH	RB	9/23/12	Reference Only	1 of 1

Water-maker Operation

Self contained Watermakers

- Most manufactures produce self contained units that house all the parts of the watermaker between the thru hull and the water tanks .

- These setups are easily controlled but require a large amount of space .

- A typical unit has sizes 8 to 10 GPH up to 83 to 85 GPH

Portable and handheld Watermakers

Katadyn Swiss portable **watermakers**, this one is recommended for liferafts. Also look at their low power consumption and 12V power **watermakers**

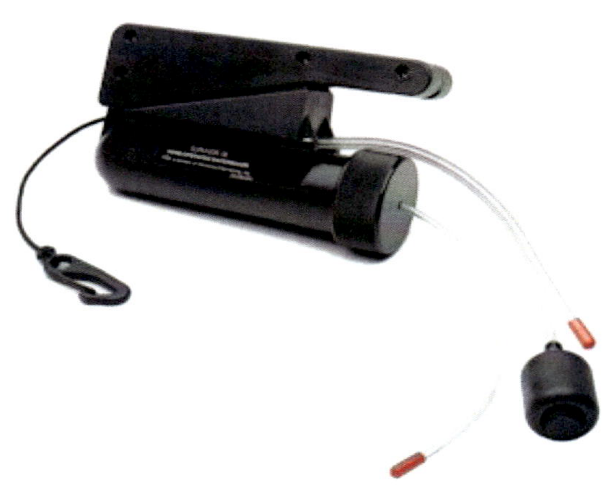

The Thru-Hull

- It is important to have good water flow through the thru hull fitting, by having a forward facing scoop type 3/4 inch thru hull as low as possible in the boat .

- Any air getting in the system will need bleeding. The intake should not be near heads or grey water outlets

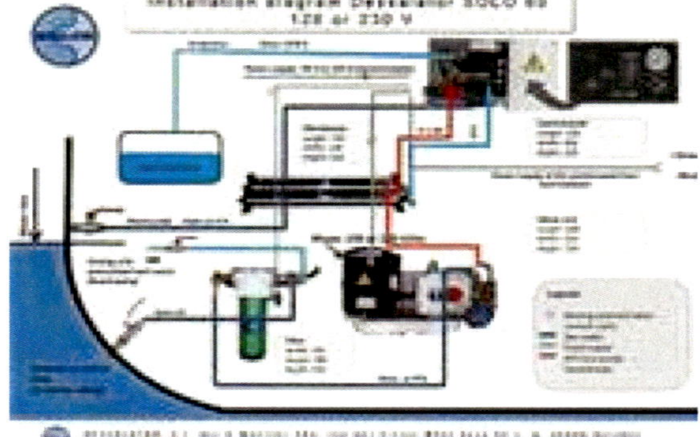

The Strainer

- A typical strainer will remove large to medium size debris that comes in with the sea water

- The strainer needs to be inspected so it needs to be accessible. Do not install near electrical equipment as water can spill during strainer cleaning

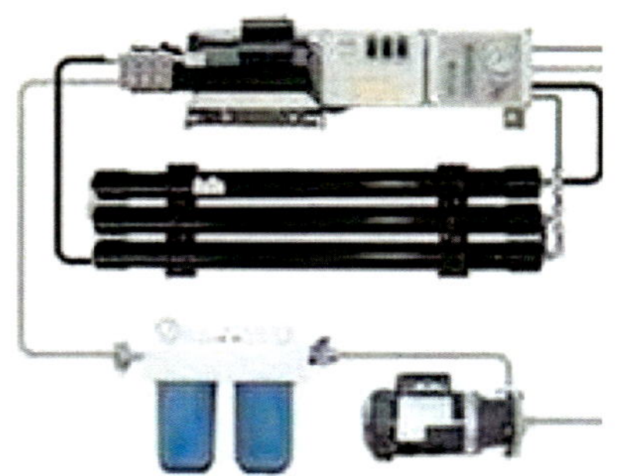 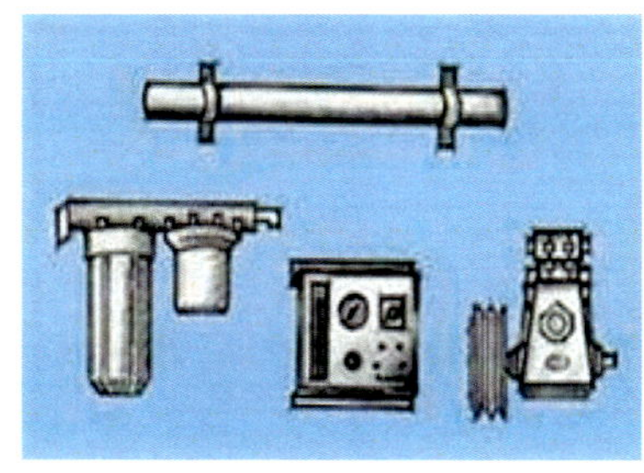 

Particle Filters

- The raw water-supply system must remove any particulate matter that could damage the high-pressure pump .

- Most systems use two filters, which are fitted with progressively finer filter elements. 30 micron and 5 micron filters are typical

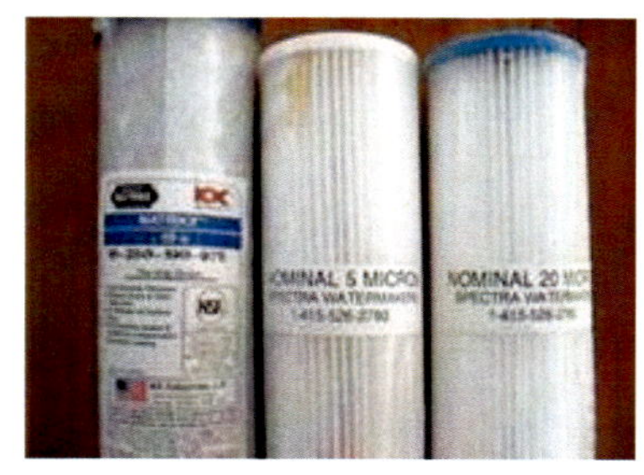 

The Low Pressure Pump

The low-pressure feed pump for water is to ensure that the high-pressure pump is always supplied with an adequate flow of water. It should also be mounted low in the boat

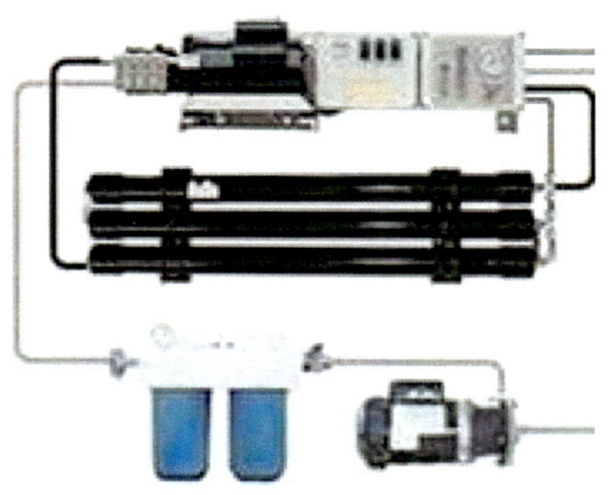

The High Pressure Pump

This pump needs to build up the pressure to around 800 psi to push the water through the membranes. The pump requires a lot of power to produce the 800 plus psi

The High Pressure Pump

Two different types of pumps are used in sailboat RO systems: electric-motor-driven plunger pumps similar to those used in pressure washers, but marinised, and hydraulic amplifiers, pumps that amplify relatively low-pressure seawater and obtain the 800 plus psi pressure required for successful RO operation

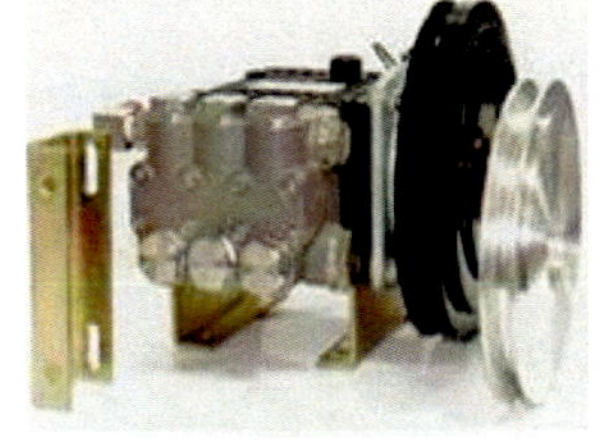

Membrane Units

The key element of the RO system, the salt-separation membrane, is semi-permeable, its small pores removing anything other than water molecules

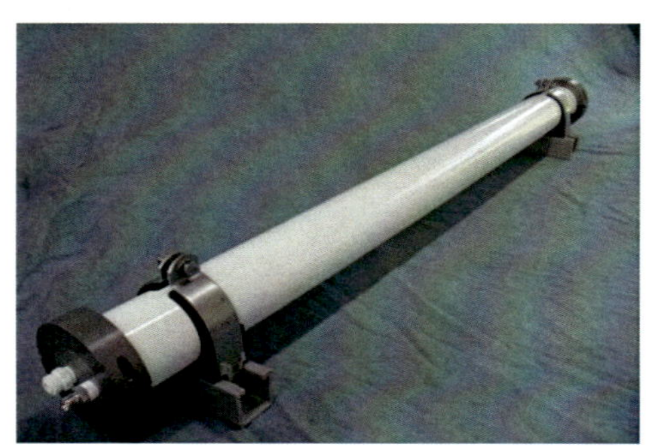

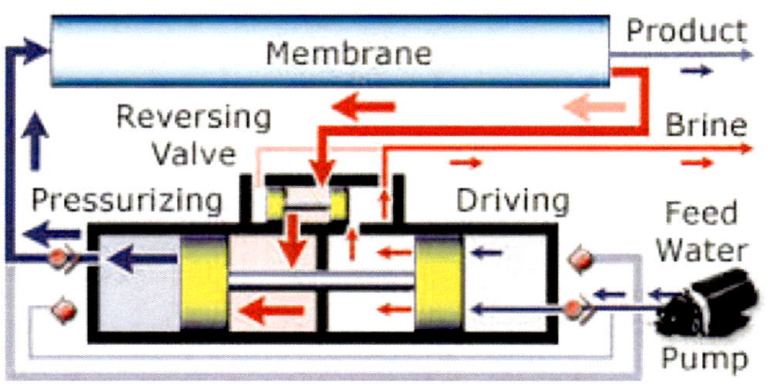

Membrane Units

- The membranes used in seawater RO systems are usually made of a TFC (thin film composite) membrane consisting of three layers, of spiral wound sheets of PA polyamide.

- The membrane needs to be flushed regularly to remove the particles and salt that have been built up on the supply side of the membrane during the RO process

Power Supply

- Power to the pumps, is supplied by battery, generator or engine driven pumps

- Large capacity watermakers generally need an AC power source. Smaller capacity and low energy watermakers can use DC AC or engine driven pumps

AC Power Supply

- Large-capacity systems are typically powered with 120- or 220-volt-AC motors and can require generators capable of delivering at least 3 to 4 kilowatts .

- On a boat with a genset which is run often or continuously a relatively small watermaker can produce the required water

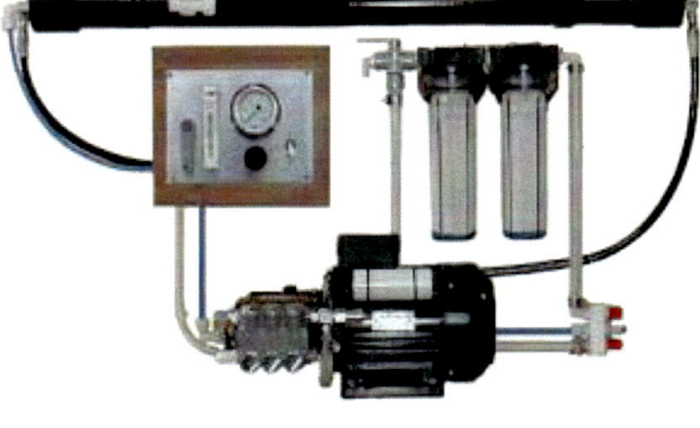

DC (12/24VDC) Power Supply

- Running the DC watermaker at the same time as the batteries are being charged will help output of the watermaker

- Make sure when you are researching a DC watermaker that the DC system on your boat can handle the loads

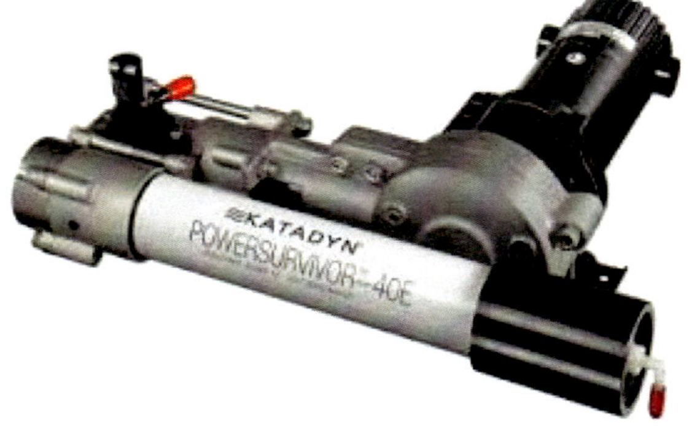

Size Matters

- How much water are you going to need, what size watermaker do you need?
- Calculate how much water you need by estimating your daily usage/person/day
- Nowadays with modern conveniences its more like 2 1/2 to 3 galls/person/day

Size Matters

- Work out how much you want to run the watermaker
- You don't want to run it all day, but you may want to run it while the generator is on, while the engine is on and you are charging your batteries
- So maybe 2-3 hour per day run time is reasonable

Maintenance Procedures

- **Filters** : Regularly check the sea water strainer and the two particle filters
- **Membrane:** Watermakers like to run often even daily for a couple of hours. Running the watermaker like this coupled with automatic back flush, which takes the minerals salt etc off the membrane surface, keeps the membrane clean and in good working order .
- A membrane not cleaned in a week can get bacterial fouling

Maintenance Procedures

Membrane: If you do not use the watermaker in a week you should flush the system or put it into storage mode which involves pickling. Membranes should not be stored dry

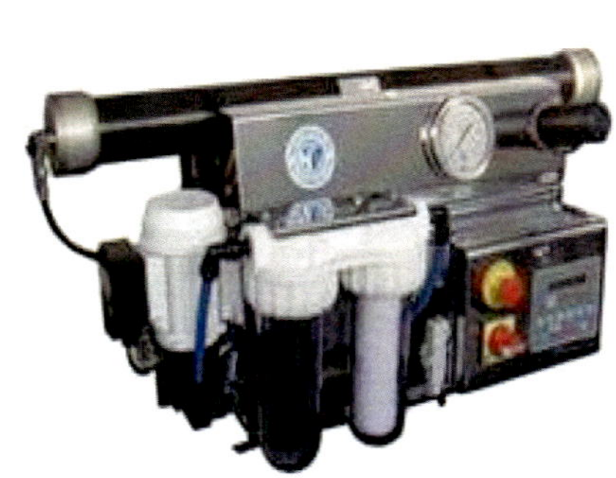

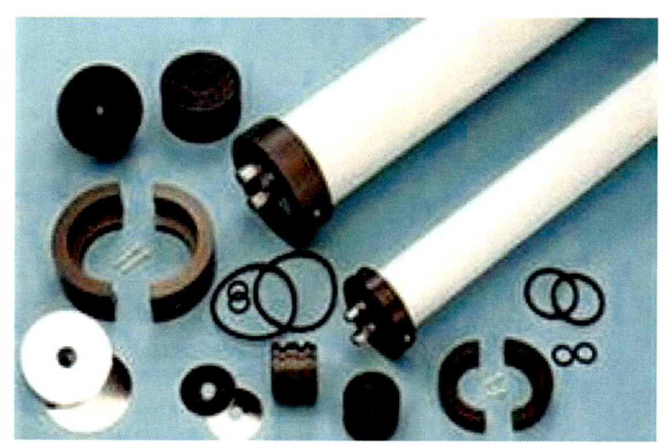

Membrane Flushing

- Some systems have an automatic flush system built in, other smaller units require manual flushing
- Some systems may have a separate tank for water for flushing

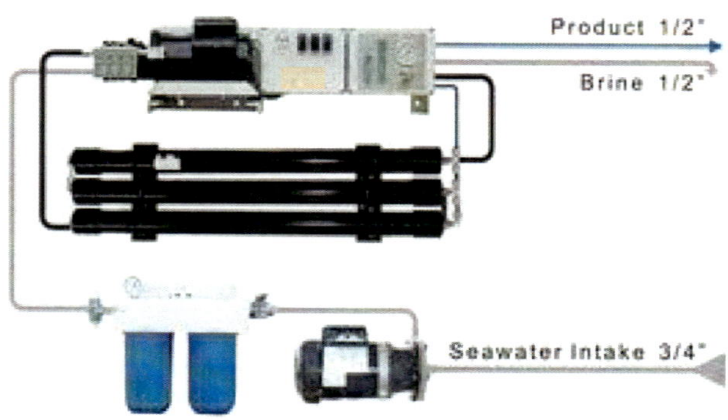

Membrane Flushing

- Manual flushing of the membrane requires switching the valves on either side of the membrane, the inlet side and the discharge side, and then turning the unit on to circulate clean water
- Flushing water needs to be free of chlorine and so a carbon filter is recommended between the tank containing the flushing water and the membrane. Chlorine can react with some cleaning chemicals

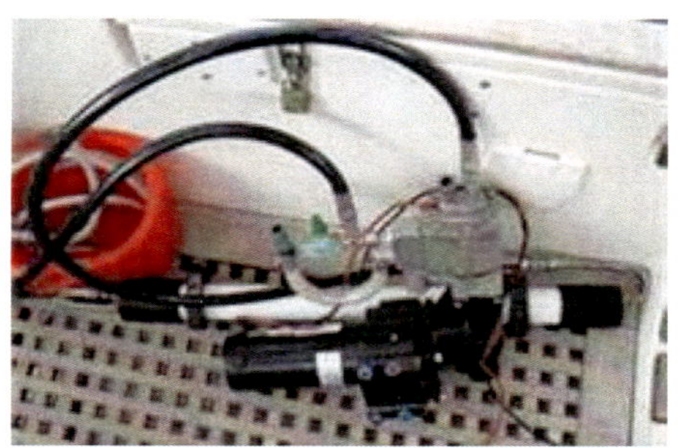

Type I/II Suction Toilet
Below Water Line

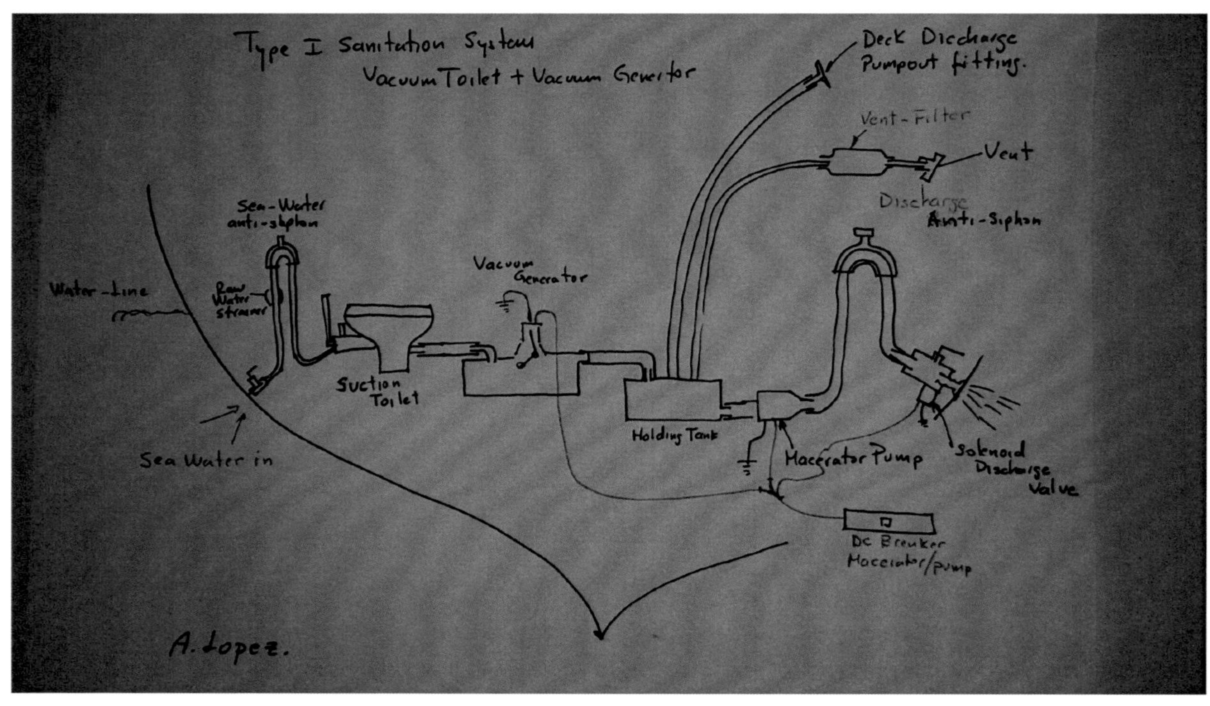

Vacuum Toilet and Vacuum Generator

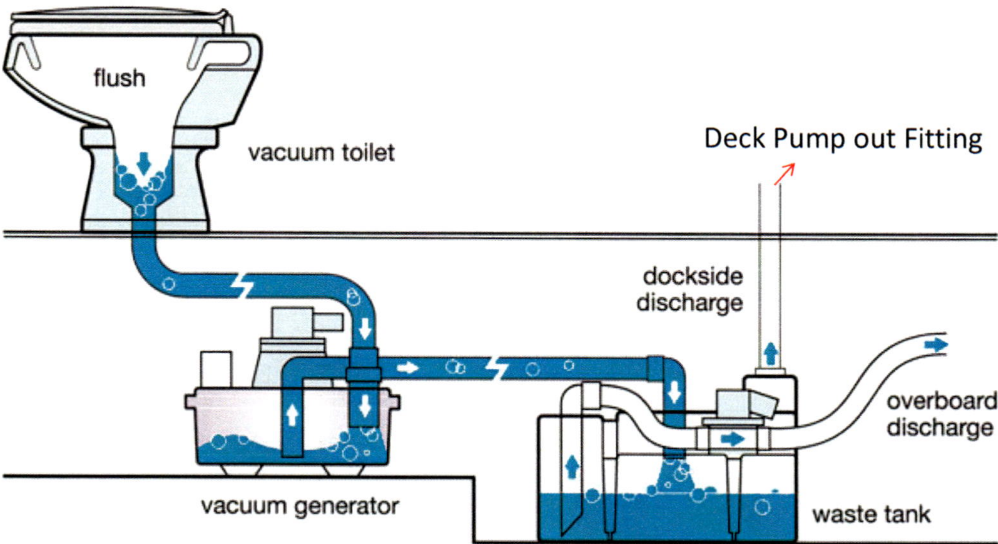

Vacuum-type marine sanitation devices typically demand extra space to accommodate the three main components: the toilet itself, the vacuum generator, and the holding tank.

Vacuum Toilet & Vacuum Generator

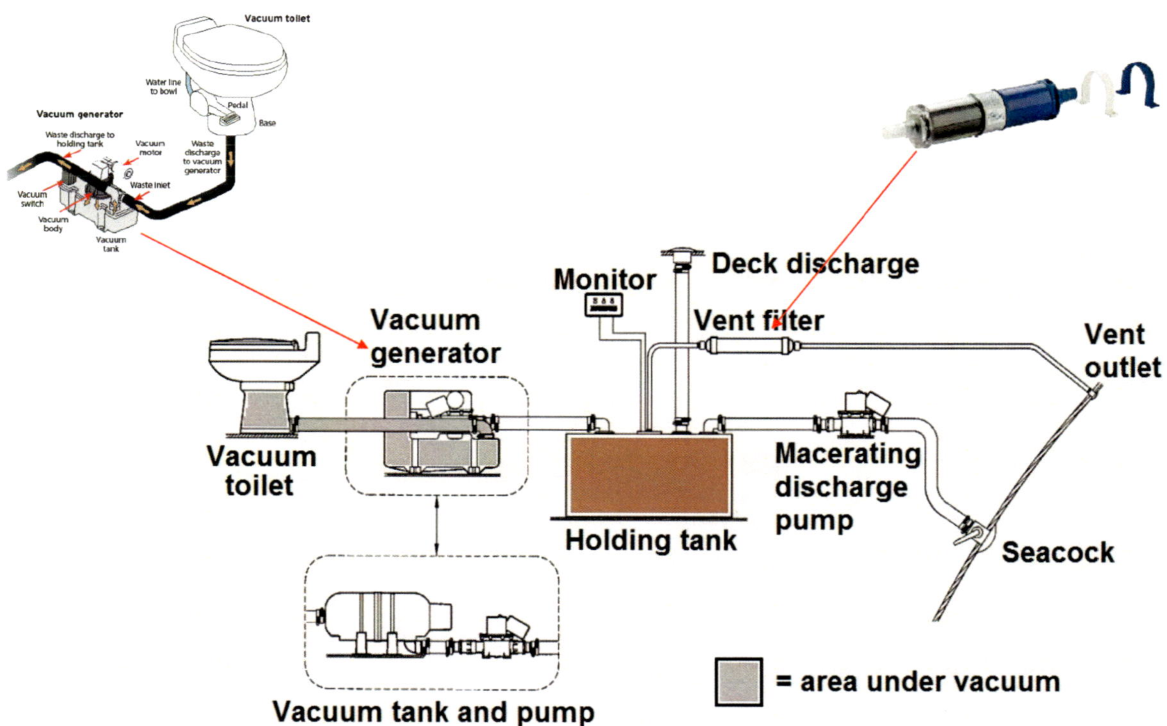

Type I Treatment (Chemical)

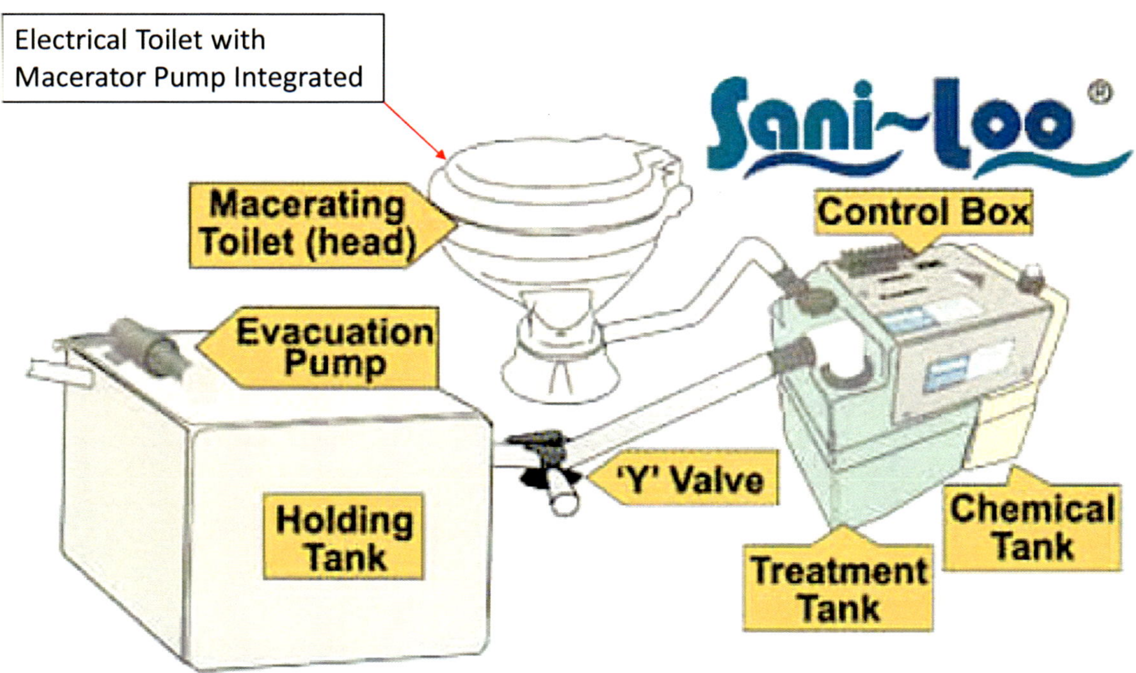

Electro-Scan Treatment Tank
Type I

The electro scan utilizes electrodes which temporarily convert salt water into a powerful bactericide. The treated waste water then safely and conveniently reverts back to its original state of salt and water

Electro-Scan Treatment Tank
Type II with Holding Tank

The design of the treatment tank optimizes the thoroughness of treatment by forcing the bactericide and bacteria into direct contact, effectively killing viruses and bacteria

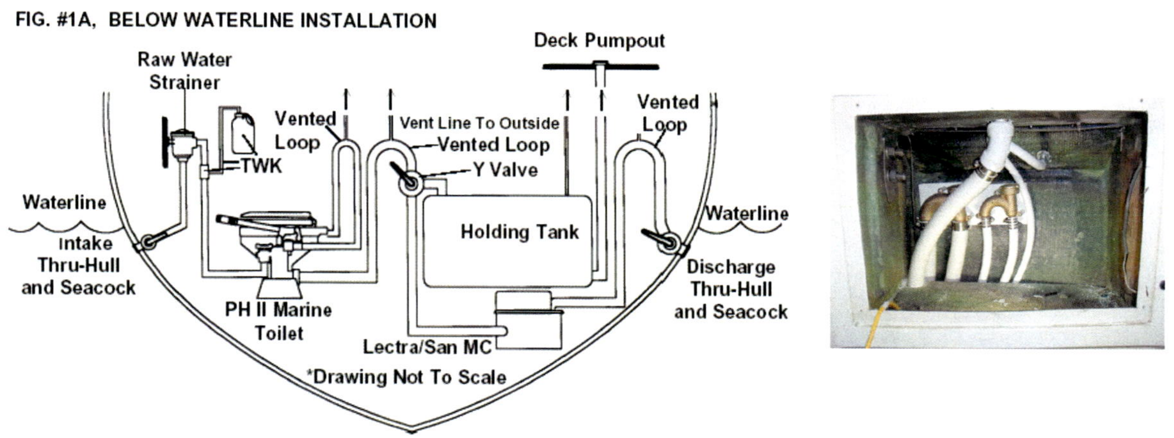

Electro-Scan Type II System
Without Holding Tank

Additionally the system could be installed above the water line and without holding tank , then the electro scan can be installed with one or two marine toilets to form a complete sanitation system. Pressing the "START STOP" pad activates the treatment cycle

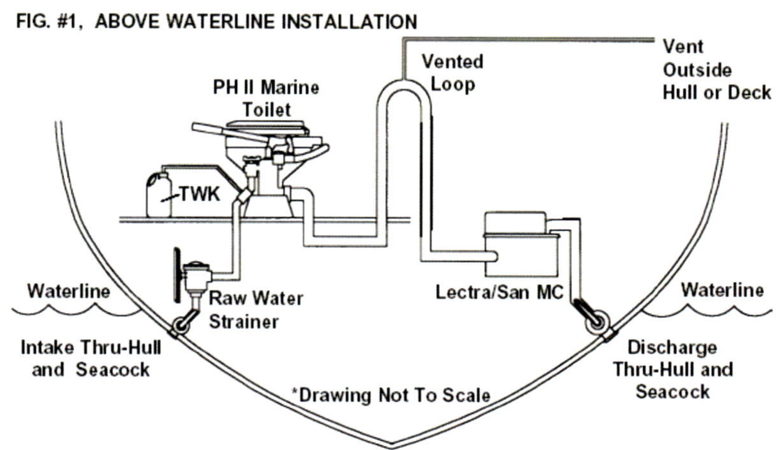

Vent Filter Location

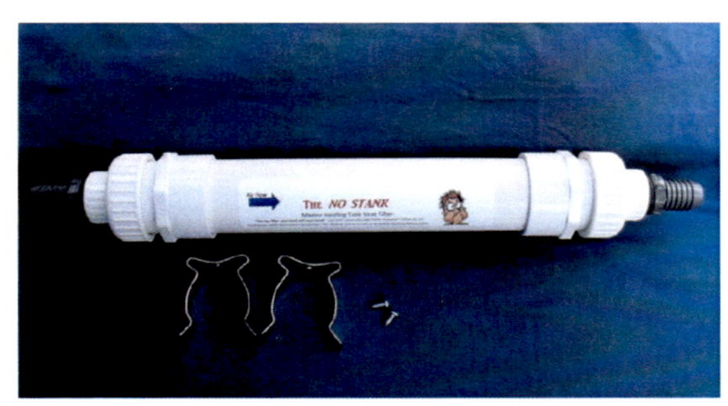

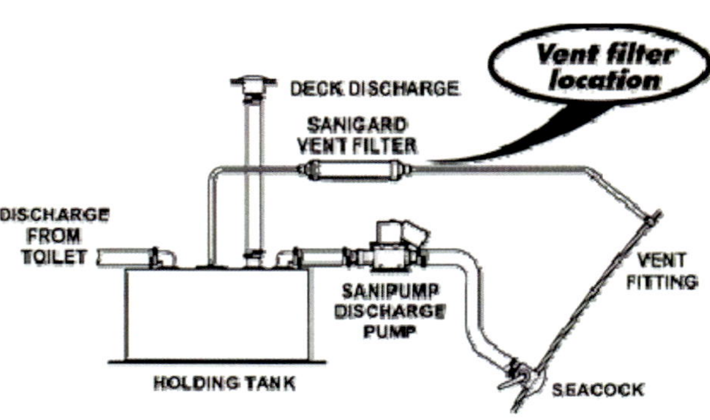

Electric Toilet & Electro-Scan

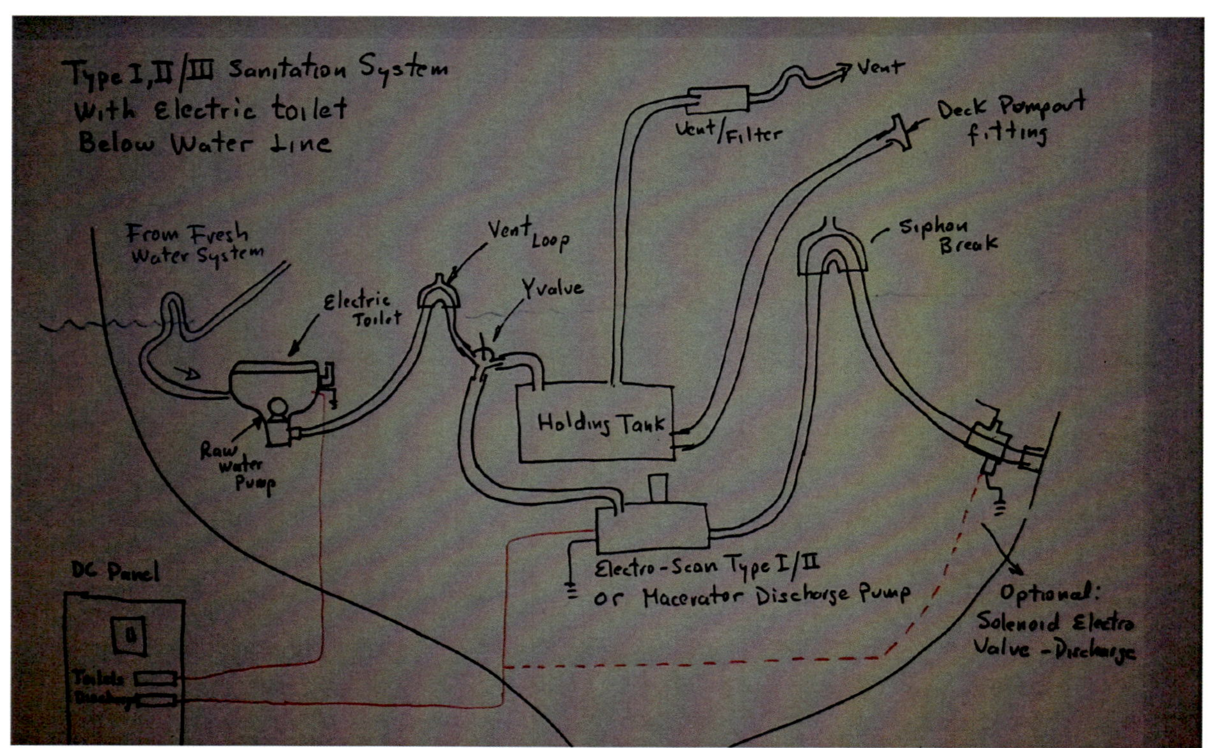

Holding Tank Evacuation

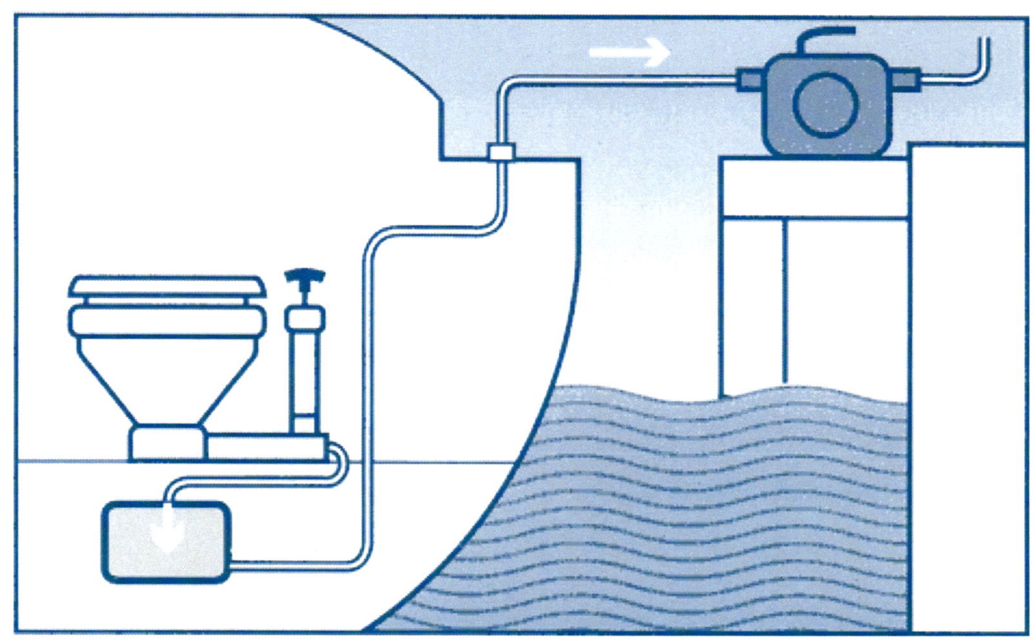

Tank Monitor

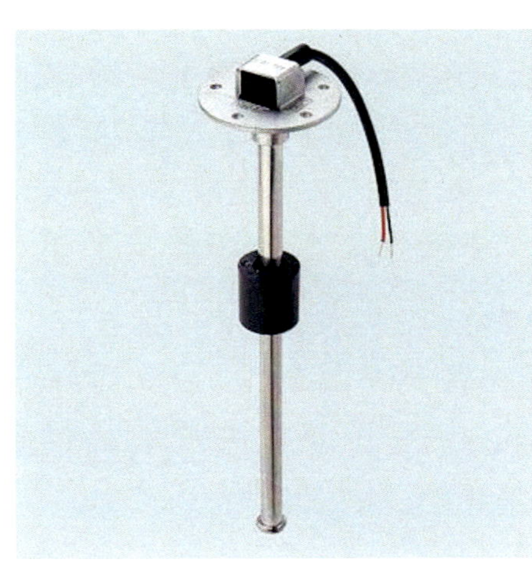

Besides onboard waste, this tank monitor also tracks
the content of graywater and potable water tanks.

A/C With Pump Relay

A/C With Pump Relay